T0135055

Computational Biology

Volume 24

The *Computational Biology* series publishes the very latest, high-quality research devoted to specific issues in computer-assisted analysis of biological data. The main emphasis is on current scientific developments and innovative techniques in computational biology (bioinformatics), bringing to light methods from mathematics, statistics and computer science that directly address biological problems currently under investigation.

The series offers publications that present the state-of-the-art regarding the problems in question; show computational biology/bioinformatics methods at work; and finally discuss anticipated demands regarding developments in future methodology. Titles can range from focused monographs, to undergraduate and graduate textbooks, and professional text/reference works.

More information about this series at http://www.springer.com/series/5769

Sourav S. Bhowmick · Boon-Siew Seah

Summarizing Biological Networks

Sourav S. Bhowmick
School of Computer Science
and Engineering
Nanyang Technological University
Singapore
Singapore

Boon-Siew Seah
School of Computer Science
and Engineering
Nanyang Technological University
Singapore
Singapore

ISSN 1568-2684
Computational Biology
ISBN 978-3-319-85438-0 ISBN 978-3-319-54621-6 (eBook)
DOI 10.1007/978-3-319-54621-6

Printed on acid-free paper

This Springer imprint is published by Springer Nature
The registered company is Springer International Publishing AG
The registered company address is: Gewerbestrasse 11, 6330 Cham, Switzerland

To our parents and wives.

Preface

Data do not give up their secrets easily. They must be tortured to confess.

Jeff Hopper, Bell Labs

The desire to study biology from a systems perspective has led to an emergence of new science—biological network analysis. Biological network models biological entities (e.g., proteins and genes) and their relationships (e.g., physical and genetic interactions) to characterize their cooperative activity within a system. With the rapid growth of such network data, the information overload problem has become a major stumbling block to analyze these networks, making human interpretation of such data increasingly difficult. Hence, there is a growing need to construct methods for large-scale topological and functional summaries of biological networks to understand the underlying mechanics of biological systems.

This book presents frameworks, as they stand today, that allow biologists to rapidly visualize and comprehend high-level topological and functional summary of the processes that govern biological systems via topological or functional organization *within* a biological network (intra-system processes) and relationships *between* biological networks (inter-system processes). Drawing on well-founded principles in data mining, systems biology, and bioinformatics, we present a multi-resolution and multi-perspective analysis paradigm to address this broad goal. Note that it is reasonable to expect this picture to change with time.

As a representative example of biological networks, we utilize protein–protein interaction (PPI) networks in majority part of this book. Our discussion is divided into five parts. First, we have attempted to review, as accurately as possible, a wide spectrum of approaches proposed by the bioinformatics community to cluster PPI networks and highlight their strengths and limitations. The results of such clustering can be considered as a summary of topological or functional modules in the underlying PPI network. In particular, a pervasive desire of this review is to emphasize the uniqueness of the network clustering problem in the context of PPI networks and highlight why a panoply of generic network clustering algorithms proposed by the data mining community cannot be leveraged to address this problem effectively.

Second, we review a closely related problem to PPI network clustering, functional summarization, which can enable us to make sense out of the information contained in large PPI networks by generating multi-level functional summaries. We discuss a data-driven and generic PPI network summarization framework that constructs higher level functional summary to summarize the underlying PPI network to obtain a concise, interpretable representation of the network. It generates the "best" summary from both interaction and annotation data by maximizing information gain for a specific resolution. We evaluate the performance of this framework on several real-world PPI networks, its superiority over network clustering, and showcase its applicability in comprehending Alzheimer's disease network.

Third, we discuss a technique that summarizes a PPI network in a multi-perspective manner. This is based on the fact that a biological system can be seen from different functional perspectives (e.g., components in a PPI network can be organized by localization, process, disease, etc.). Each discovered perspective represents a distinct interpretation of how the network can be functionally summarized. The performance of this framework is extensively discussed with several real-world PPI networks highlighting the limitations of network clustering paradigm to generate such multi-perspective summary. We also performed a case study using human autophagy system to illustrate the utility of this framework.

Fourth, we discuss a data-driven effort to construct summaries of differential functional responses of gene interaction networks that undergo "rewiring" after environmental change. Experimental evaluation with real-world dataset demonstrates the superiority of this technique to address the differential network summarization problem.

The last topic consists of several open problems of this young field. The list presented should by no means be considered exhaustive and is centered around challenges and issues currently in vogue. Nevertheless, readers can benefit by exploring the research directions given in this part.

The book is suitable for use in advanced undergraduate- and graduate-level courses on biological networks. It has sufficient material that can be covered as part of a semester-long course, thereby leaving plenty of room for an instructor to choose topics. An undergraduate course in algorithms, graph theory, and basic cell biology should suffice as a prerequisite for most of the chapters. A good knowledge of C++/Java programming language is sufficient to code the algorithms described herein. For completeness, we have provided background information on several topics in Chap. 2: the central dogma of biology, protein–protein interactions, high-throughput experimental techniques to analyze protein–protein interactions, and annotations of these interactions with Gene Ontology. The knowledgeable reader may omit this chapter and perhaps refer back to comparisons while reading later chapters of this book.

We hope that this book will serve as a catalyst in helping this burgeoning area of biological network summarization grow and have practical impact.

Singapore
December 2016

Sourav S. Bhowmick
Boon-Siew Seah

Acknowledgements

It is a great pleasure for us to acknowledge the assistance and contributions of a large number of individuals to this effort. First, we would like to thank our publisher Springer-Verlag for their support. In particular, we would like to acknowledge the efforts, help, and patience of Melissa Fearon and Jennifer Malat, our primary contacts for this edition.

The work reported in this book grew out of the PANORAMA project at the Nanyang Technological University (NTU), Singapore, and Massachusetts Institute of Technology (MIT), USA, under the auspices of Singapore-MIT Alliance graduate program. In this project, we explored issues on building frameworks that allow biologists to rapidly visualize the processes that govern biological systems. Specifically, the chapters in this book are part of Boon Siew's thesis work under the guidance of Sourav. Some of these chapters are published in reputable journals and conferences in the area of bioinformatics and systems biology.

Dr. C. Forbes Dewey, Jr. of MIT, who was a key collaborator for this project, deserves the first thank you. Not only did he introduce us to interesting and exciting field of network biology and network medicine, but he was also always willing to discuss ideas with us, no matter how strange they were.

In addition, we would also like to express our gratitude to all the group members and collaborators, past and present, in our **Co**mputational **S**ystems **B**iolog**y** research group (COSBY). In particular, Dr. Huey Eng Chua (NTU), Dr. Jie Zheng (NTU), Dr. Lisa Tucker-Kellogg (Duke-NUS Medical School), Dr. Hanry Yu (NUS), and Mengxuan Chen made substantial contributions to the broader aspect of our research in network biology.

Quite a few people have helped us with the initial vetting of the text for this book. It is our pleasure to acknowledge them all here. We would like to thank Scientific Publishing Services (SPS) for carefully proofreading the complete book in a short span of time and suggesting the changes which have been incorporated.

Sourav and Boon Siew would like to acknowledge their parents and family members who gave them incredible support throughout the years. A special thanks goes to Dr. Paul Matsudaira (MIT, NUS) and Dr. Hew Choy Leong (NUS), who were a great motivator during our early days when we were grappling with the new field of systems biology. They were the major force behind our continuous strive for breaking out from the comfort zone of computer science to explore problems that are at the intersection of two or more disparate fields. It has been a great learning experience for us.

Finally, we would like to thank the Singapore-MIT Alliance for the generous resources and financial support provided for the PANORAMA project. We would also like to thank the School of Computer Science and Engineering at the Nanyang Technological University for allowing the use of their resources to help complete this book.

Singapore
December 2016

Sourav S. Bhowmick
Boon-Siew Seah

Contents

Chapter 1
Introduction

For decades, scientists have studied the components of living systems in isolation [1]. For instance, in a study of proteins, genes, or even biological pathways, the component of interest is first removed from its true environment and then studied by observing its individual properties. This approach has served the research community well, and has been – and still is – an extremely effective technique at uncovering the properties of molecular components (components for brevity) at a detailed level. However, its limitations are also apparent in recent times [2]. While effective at studying their behavior and properties in isolation, the behavior of isolated molecular components often cannot be trivially extrapolated to groups of components when put together. For instance, the behavior of proteins *in vitro* often contradicts the behavior *in vivo* [1]. Proteins often play multiple roles (*moonlighting*), and the processes in which they take part are contextual and dynamic [3]. Even biological processes themselves do not operate in isolation; instead, they are a well orchestrated cooperation among multiple processes. An extreme example is that of social organisms (e.g., ants) and their social structure needed to survive and operate together [4].

In light of this, the approach of viewing biological systems from a broader, global perspective is an increasingly attractive enterprise [2, 5, 6]. Rather than modeling components in an isolated, reductionist manner, the cooperative activity of a group of components is modeled as well. This "systems-based" paradigm looks at not just the individual components, but also their activity and relationships as a cooperative whole [5]. The most well-known method to model biological systems in this manner is through *biological networks* (graphs) [7].

A biological network is modeled as a graph $G = (V, E, w)$, where V is a set containing the components of the network, $E \subseteq V \times V$ is a set containing the pairwise relationships between the components, and $w : E \rightarrow \mathbb{R}$ is a real-valued weight function that assigns weights to each $e \in E$. It lays out the structure of the

S.S. Bhowmick and B.-S. Seah, *Summarizing Biological Networks*,
Computational Biology 24, DOI 10.1007/978-3-319-54621-6_1

components and their relationship and enables mathematical analysis to be performed on this structure. The most common class of biological networks is the *protein-protein interaction network* (PPI network) [7]. Here, V is a set of proteins in the PPI network to be modeled, E is the set of physical interactions among the proteins in V, and w is a function that models the strength or confidence of the interactions. Another example is a biological network model of pathway-pathway interactions [8]. One toy example of a pathway-pathway interaction network is a network of interactions between cell-cycle, apoptosis, and DNA repair pathways. In this case, V is the set of pathways, E is the set of pathway-pathway relationships, and w maps the relationship strength between pathways. Many other classes of biological networks exist. They may range from neuronal and disease networks to mRNA networks, transcriptional regulatory networks, and DNA-protein interaction networks. Although there are unique properties that defines each class of network, many common network properties emerge among them, and consequently, many analytical methods that apply to one class of network can be transferred to other classes of networks.

The desire to study biological systems from a global perspective has led to an emergence of new science–*biological network analysis*. With network analysis, one may uncover key system-wide properties and behaviors of a biological system that reductionist methods could not. In the seminal paper by Barabasi et al. [9], the authors discovered the *scale-free* distribution of biological networks and proposed the preferential attachment model of real-world networks. Many biological networks (e.g., PPI network) demonstrate small-world property and high degree of clustering. Consequently, it contains a few highly connected hubs and there are relatively short paths between any pair of nodes in the network. Many other subsequent studies reveal other important models and properties of biological networks using networks analysis, including the party and date hub model of proteins [10] and the evolutionary models of PPI networks [11]. Despite their importance to systems level biology, there is still a chasm between networks analysis that searches for general properties and *functional analysis*[1] of networks needed by a biology researcher. While the aforementioned studies uncover general properties of proteins and their interactions, others may still wish to interpret networks in a more specific and concrete manner. We justify this with an example. Consider an analysis of the Alzheimer's Disease PPI network [12]. Network analysis may reveal interaction distribution and characteristic of Alzheimer's Disease from a general viewpoint – for example, the degree distribution of the proteins in the network – but a typical researcher studying the Alzheimer's Disease PPI network may also want to look for more concrete patterns and observations. For instance, one may wish to look for a summary of most important functional processes and their relationships (e.g., the relationship

[1]Network functional analysis is the analysis of the underlying biological roles and function of the network (and its subnetworks).

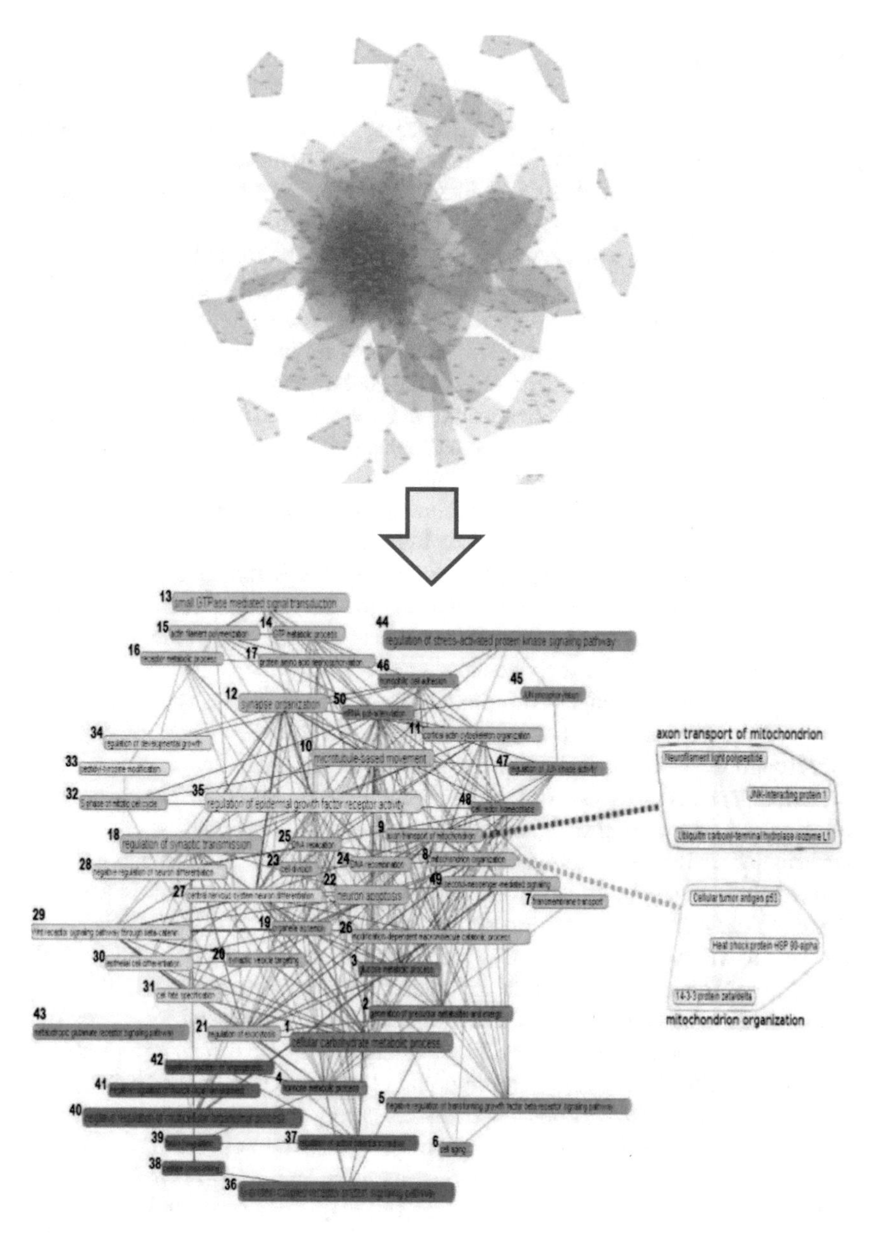

Fig. 1.1 Summary of important functional processes in chromatin PPI network

between transport and apoptosis processes) that take part in the network. Figure 1.1 illustrates another example of a summary that groups related functional processes in the chromatin PPI network.

1.1 Challenges

The complexity of modeling biological processes from a systems viewpoint gives rise to several challenges. The first challenge associated with biological network analysis is the amount of data it needs to deal with. The information needed to perform a global perspective analysis is daunting. To illustrate this, consider a simple case of 100 proteins in a biological system. While there are only 100 proteins to study in isolation, the networks approach studies not just the 100 proteins, but also potentially 10,000 pairwise relationships among them. Even in this simple case, the combinatorial complexity of network analysis literally increases the complexity of the study by several orders of magnitude. If the cost of acquiring 10,000 relationship data is prohibitive, then systems-based study is not even close to feasible. Fortunately, recent advances in high-throughput technologies (e.g., *yeast-two-hybrid* [13]) have played a massive role in enabling such studies [14].

The second important challenge is noisiness of the data. A natural consequence of large scale automated techniques like two-hybrid screening is the high rate of false positives and false negatives [15]. It is important for any analytical tool to take noise into account to guard against spurious predictions. The approach proposed in [16], for example, takes into account the noisy nature of high-throughput data in order to predict interactions from heterogeneous sources.

Finally, the third challenge is information overload, which is a byproduct of dealing with large volumes of interaction data. Specifically, the deluge of data from high-throughput experiments comes at a cost. A biologist may find data provided by interaction datasets in its raw form overwhelming (e.g., Fig. 1.2). The difficulty of analyzing and interpreting such complex dataset is called *information overload*. As such, a major challenge to biologists is to make sense out of the intertwining hairball of information contained in large biological networks. One may wish to find ways to extract summarized information about a biological network. Alternatively, one may wish to find ways to compare several biological networks to identify significant patterns. This may allow one to distinguish regions of the network that undergo significant changes in its diseased state compared to its normal state.

Given the above challenges, a multitude of algorithms have been proposed in the literature [17–20]. We highlight an important class of such network analysis algorithms that is relevant to this book – network clustering and summarization. Network clustering aims to identify densely interacting regions of a network. The clustering process assists in summarizing a biological network and also to reveal interesting functional predictions regarding the cluster. In [21], network clustering is applied on the global yeast PPI network to uncover the landscape of important *functional modules* within the network including protein complexes. In this book, we shall discuss techniques for clustering and summarizing biological networks and discuss their strengths and limitations. Recall the large number of interaction and interactor attributes are provided by biological network data. When confronted with such a deluge of data, biological researchers, for instance, are still limited in their ability to manually interpret and analyze PPI networks together with their protein attributes.

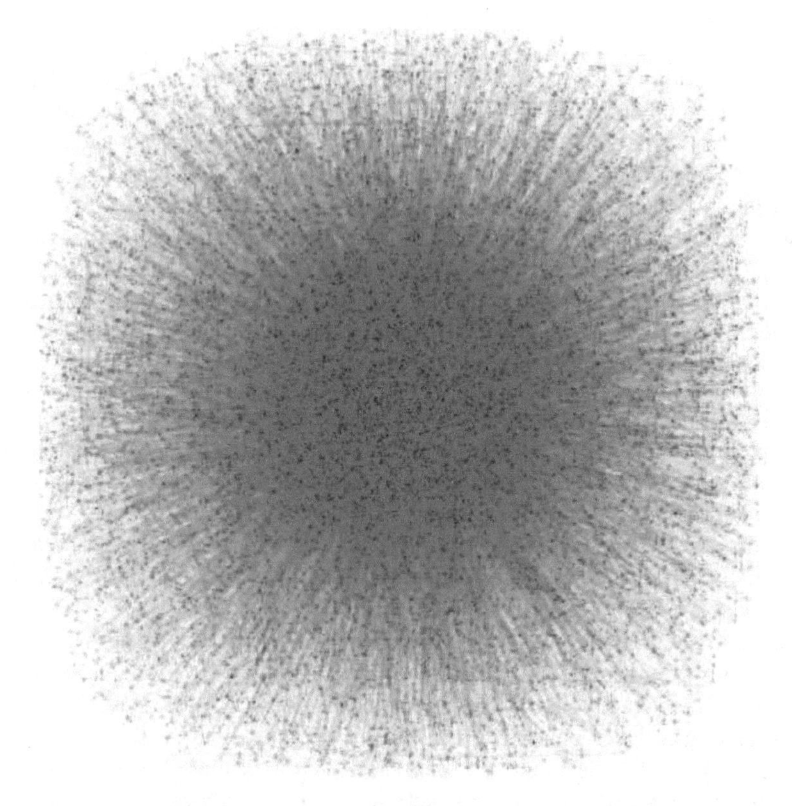

Fig. 1.2 Information overload in a human PPI network comprising more than 9000 proteins

Each protein may be annotated with hundreds, if not thousands, of attribute information. Together with the large number of proteins and their interactions, interpreting these data as whole can be a daunting task.

1.2 Overview of This Book

This book focuses on making sense out of this deluge of biological network data. Specifically, we discuss a variety of techniques that address the network clustering and summarization problems. We aim to bridge the gap between the complexity of large scale biological network data and concise interpretability demanded by a typical biologist.

In particular, a major focus of this book is on *attributed* PPI networks, which is an extension of general PPI networks. Instead of modeling the proteins in a PPI network as homogeneous entities, in attributed PPI networks we endow each component with attributes (such as functional annotations). Hence attributed PPI networks refer to networks whose proteins are not treated as homogeneous nodes, but entities that have attributes associated with it. For instance, a protein can have *Gene Ontology* (GO) term annotations as its attributes. The richness provided by this extended graphical model introduces additional challenges to their analysis (for example, the high dimensionality of protein attributes), but at the same time, it opens the door for opportunities to yield novel findings. Few studies have considered network analysis on attributed PPI networks. We shall describe later how such networks can be obtained, and how their analysis using a variety of recently proposed methods are advantageous to standard attribute-agnostic approaches. Towards addressing existing limitations, especially with regard to information overload, we organize the discussions in this book as follows.

- In Chap. 2 we present the elements that serve as background for the remaining chapters of the book. In particular, we focus on the modeling of biological systems as networks. We discuss how interaction data is acquired, and we discuss several knowledge-bases that provide a wealth of interaction and functional information. We also discuss GO and gene function annotations [22], which provide controlled annotation describing functions and activities of genes, gene products (including proteins) and their interactions.
- Chapter 3 introduces progress made in clustering PPI networks. We present a review of existing methodologies together with an evaluation of the strengths and limitations of current tools, especially with respect to their ability to assist biologists in interpreting complex biological networks.
- In Chap. 4, we discuss limitations of traditional network clustering techniques and introduce a methodology that constructs *functional summaries* of any PPI network. The central goal of this approach is to provide researchers a summary of a PPI network from a functional perspective. The summary reduces the complex "hairball" of PPI data into concise *functional subgraphs* that, along with their interactions, represent a compressed functional representation of the underlying PPI network. Unlike graph clustering algorithms that focus on finding strongly coherent subgraphs, a functional summarization technique is focused on ensuring the modules are representative of the functions described by the summary and the entire summary is representative of the underlying PPI network. Functional summarization allows researchers to overcome information overload problem associated with the interpretation of large-scale PPI networks, allowing visual interpretation of the functional components and their interactions that underlie a PPI network. In addition, it provides an ability to control the granularity of this summary, with which we can construct multiple layers of "bird's-eye" view summaries with varying complexity.
- Chapter 5 discusses a technique that addresses the limitation of existing graph clustering methods, which present only *one* clustering perspective of a PPI network. Instead, it recognizes that most PPI networks can be organized in multiple ways. To this end, it presents a recent technique that extends the capability of network summarization techniques to construct an atlas of summaries of the underlying PPI network. Each summary represents a *facet* of the network that represents a

particular functional organization, and the set of summaries presented are *functionally orthogonal*. Intuitively, functionally orthogonal summaries are a collection of summaries, such that each summary represents a unique functional organization of the network that is different from the rest of the summaries.

- Recently, [23] proposed a technique to construct a *differential network* (dE-MAP *network*) from two static gene interaction networks (i.e., E-MAP network) in order to map the interaction differences between them under environment or condition change (e.g., DNA-damaging agent). This differential network is then manually analyzed to conclude that DNA repair is differentially effected by the condition change. Unfortunately, manual construction of *differential functional summary* from a dE-MAP network that summarizes all pertinent functional responses is time-consuming, laborious and error-prone, impeding large-scale analysis on it. In Chap. 6, we present an automated technique that summarizes pertinent functional differences between two E-MAP networks under contrasting conditions by leveraging GO annotations in order to obtain a high-level map of functional responses due to condition change.
- Finally, in Chap. 7 we summarize the contributions of this book and list down interesting open research problems in this arena of clustering and summarizing biological networks.

References

1. R.T. Peterson, Chemical biology and the limits of reductionism. Nature Chem. Biol. **4**, 635–638 (2008)
2. A.C. Ahn, M. Tewari, C.-S. Poon, R.S. Phillips, The limits of reductionism in medicine: could systems biology offer an alternative? PLoS Med. **3**, e208 (2006)
3. C.J. Jeffery, Moonlighting proteins: old proteins learning new tricks. Trends Genet. TIG **19**, 415–417 (2003)
4. C.R. Smith, A. Dolezal, D. Eliyahu, C.T. Holbrook, J. Gadau, Ants (Formicidae): models for social complexity. Cold Spring Harb. Protoc., **7**, pdb.emo125 (2009)
5. T. Ideker, T. Galitski, L. Hood, A new approach to decoding life: systems biology. Ann. Rev. Genomics Hum. Genet. **2**, 343–372 (2001)
6. U. Sauer, M. Heinemann, N. Zamboni, Genetics. Getting closer to the whole picture. Science (New York) **316**, 550–551 (2007)
7. U. Alon, Biological networks: the tinkerer as an engineer. Science (New York, N.Y.) **301**, 1866–1867 (2003)
8. D. Dotan-Cohen, S. Letovsky, A.A. Melkman, S. Kasif, Biological process linkage networks. PloS One **4**, e5313 (2009)
9. A. Barabasi, R. Albert, Emergence of scaling in random networks. Science (New York, N.Y.) **286**, 509–512 (1999)
10. D. Ekman, S. Light, A.K. Björklund, A. Elofsson, What properties characterize the hub proteins of the protein-protein interaction network of Saccharomyces cerevisiae? Genome Biol. **7**, R45 (2006)
11. T. Yamada, P. Bork, Evolution of biomolecular networks: lessons from metabolic and protein interactions. Nat. Rev. Mol. Cell Biol. **10**, 791–803 (2009)
12. S. Kerrien, Y. Alam-Faruque, B. Aranda, I. Bancarz, A. Bridge, C. Derow, E. Dimmer, M. Feuermann, A. Friedrichsen, R. Huntley, C. Kohler, J. Khadake, C. Leroy, A. Liban, C. Lieftink, L.

Montecchi-Palazzi, S. Orchard, J. Risse, K. Robbe, B. Roechert, D. Thorneycroft, Y. Zhang, R. Apweiler, H. Hermjakob, IntAct-open source resource for molecular interaction data. Nucleic Acids Res. **35**, D561–D565 (2007)

13. T. Ito, T. Chiba, R. Ozawa, M. Yoshida, M. Hattori, Y. Sakaki, A comprehensive two-hybrid analysis to explore the yeast protein interactome. Proc. Natl. Acad. Sci. **98**, 4569–4574 (2001)
14. P. Uetz, L. Giot, G. Cagney, T.A. Mansfield, R.S. Judson, J.R. Knight, D. Lockshon, V. Narayan, M. Srinivasan, P. Pochart, A. Qureshi-Emili, Y. Li, B. Godwin, D. Conover, T. Kalbfleisch, G. Vijayadamodar, M. Yang, M. Johnston, S. Fields, J.M. Rothberg, A comprehensive analysis of protein-protein interactions in Saccharomyces cerevisiae. Nature **403**, 623–627 (2000)
15. P. Braun, M. Tasan, M. Dreze, M. Barrios-Rodiles, I. Lemmens, H. Yu, J.M. Sahalie, R.R. Murray, L. Roncari, A.-S. de Smet, K. Venkatesan, J.-F. Rual, J. Vandenhaute, M.E. Cusick, T. Pawson, D.E. Hill, J. Tavernier, J.L. Wrana, F.P. Roth, M. Vidal, An experimentally derived confidence score for binary protein-protein interactions. Nat. Methods **6**, 91–97 (2009)
16. I. Iossifov, M. Krauthammer, C. Friedman, V. Hatzivassiloglou, J. S. Bader, K.P. White, A. Rzhetsky, Probabilistic inference of molecular networks from noisy data sources. *Bioinformatics (Oxford, England)*, **20**, 1205–1213 (2004)
17. G.D. Bader, C.W.V. Hogue, An automated method for finding molecular complexes in large protein interaction networks. BMC Bioinf. **4**, 2 (2003)
18. A.J. Enright, S. Van Dongen, C.A. Ouzounis, An efficient algorithm for large-scale detection of protein families. Nucleic Acids Res. **30**, 1575–1584 (2002)
19. M. Kalaev, M. Smoot, T. Ideker, R. Sharan, NetworkBLAST: comparative analysis of protein networks. *Bioinformatics (Oxford, England)*, **24**, 594–596 (2008)
20. R. Singh, J. Xu, B. Berger, Global alignment of multiple protein interaction networks with application to functional orthology detection. Proc. Natl. Acad. Sci. **105**, 12763–12768 (2008)
21. N.J. Krogan, G. Cagney, H. Yu, G. Zhong, X. Guo, A. Ignatchenko, J. Li, S. Pu, N. Datta, A.P. Tikuisis, T. Punna, J.M. Peregrín-Alvarez, M. Shales, X. Zhang, M. Davey, M.D. Robinson, A. Paccanaro, J.E. Bray, A. Sheung, B. Beattie, D.P. Richards, V. Canadien, A. Lalev, F. Mena, P. Wong, A. Starostine, M.M. Canete, J. Vlasblom, S. Wu, C. Orsi, S.R. Collins, S. Chandran, R. Haw, J.J. Rilstone, K. Gandi, N.J. Thompson, G. Musso, P. Onge, S. Ghanny, M.H.Y. Lam, G. Butland, A.M. Altaf-Ul, S. Kanaya, A. Shilatifard, E. O'Shea, J.S. Weissman, C.J. Ingles, T.R. Hughes, J. Parkinson, M. Gerstein, S.J. Wodak, A. Emili, J.F. Greenblatt, Global landscape of protein complexes in the yeast Saccharomyces cerevisiae. Nature **440**, 637–643 (2006)
22. M.A. Harris, J. Clark, A. Ireland, J. Lomax, M. Ashburner, R. Foulger, K. Eilbeck, S. Lewis, B. Marshall, C. Mungall, J. Richter, G.M. Rubin, J.A. Blake, C. Bult, M. Dolan, H. Drabkin, J.T. Eppig, D.P. Hill, L. Ni, M. Ringwald, R. Balakrishnan, J.M. Cherry, K.R. Christie, M.C. Costanzo, S.S. Dwight, S. Engel, D.G. Fisk, J.E. Hirschman, E.L. Hong, R.S. Nash, A. Sethuraman, C.L. Theesfeld, D. Botstein, K. Dolinski, B. Feierbach, T. Berardini, S. Mundodi, S.Y. Rhee, R. Apweiler, D. Barrell, E. Camon, E. Dimmer, V. Lee, R. Chisholm, P. Gaudet, W. Kibbe, R. Kishore, E.M. Schwarz, P. Sternberg, M. Gwinn, L. Hannick, J. Wortman, M. Berriman, V. Wood, N. de la Cruz, P. Tonellato, P. Jaiswal, T. Seigfried, R. White, The Gene Ontology (GO) database and informatics resource. Nucleic Acids Res., **32**, D258–D261 (2004)
23. S. Bandyopadhyay, M. Mehta, D. Kuo, M.-K. Sung, R. Chuang, E.J. Jaehnig, B. Bodenmiller, K. Licon, W. Copeland, M. Shales, D. Fiedler, J. Dutkowski, A. Guénolé, H. van Attikum, K.M. Shokat, R.D. Kolodner, W.-K. Huh, R. Aebersold, M.-C. Keogh, N.J. Krogan, T. Ideker, Rewiring of genetic networks in response to DNA damage. Science (New York, N.Y.) **330**, 1385–1389 (2010)

Chapter 2
Background

This chapter provides an overview of key topics that serve as background for the rest of the book. First, we discuss the roles of proteins in living organisms. This is followed by a brief discussion on protein-protein interactions and methods for analyzing them. Finally, we briefly highlight on the roles of databases, ontologies and annotations in proteins and their interactions.

2.1 Proteins: The Building Block of Life

The basic building block of all living organisms is the cell. The cell itself is a complex machinery—within it a plethora of processes and components that govern the mechanisms of the cell [23]. Microtubules, tubular shaped scaffolds of the cell, provide not only shape and structure, but also act as tracks for transporting cellular cargoes. Mitochondrions are the molecular engines of the cell, generating fuel to power cellular machines. These are just a few examples of cellular components that regulate the cell machinery.

The various parts of the cell work in tandem to regulate *biological processes*—functionalities performed within the cell that control its behavior and state, depending on its internal and external environments. For example, the cell cycle is a biological process that controls the growth and replication of itself. Transport processes cargo cellular components within the cell, as well as exporting cargoes out of the cell and importing cargoes into it. Homeostatic processes regulate the equilibrium of chemical concentration in the cell to a desirable optimum.

Remarkably, the machines that run biological processes of cells are largely performed by one class of molecules called *proteins* [30]. A protein is composed of a

S.S. Bhowmick and B.-S. Seah, *Summarizing Biological Networks*,
Computational Biology 24, DOI 10.1007/978-3-319-54621-6_2

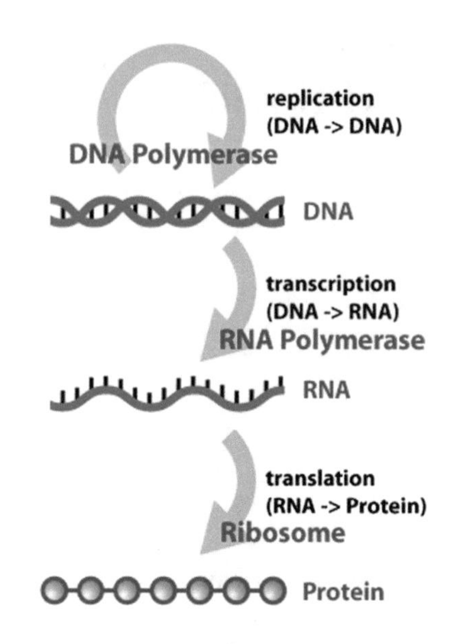

Fig. 2.1 The central dogma of molecular biology. (Image by Dhorspool at en.wikipedia)

linear sequence of amino acids that are folded into a 3D structure. Informally, one can think of proteins as strings of words formed by an alphabet of amino acids. There are 20 "canonical" amino acids in eukaryotes [36]. Each amino acid exhibits distinct chemical properties (such as polarity and hydrophobicity) and also physical properties (such as mass), giving a 3D structured protein its character and behavior. The roles of proteins are many and varied. For instance, the protein `actin` lends structural integrity to cells. Enzymes are a special class of proteins that catalyze chemical reactions. Signaling proteins like `Ras` act as messengers that amplify and distribute signals from a stimuli.

Given the significance of proteins, this begs an important question: what directs their construction and regulation? *Genetic information* is the information required for construction of proteins. The central dogma of molecular biology [7] states that genetic information flows from deoxyribonucleic acid (DNA) to oxyribonucleic acid (RNA) to protein (Fig. 2.1). Essentially, the DNA (a sequential chain of polymers called nucleotides) serves as the blueprint for the construction of proteins. The sequence of nucleotides in DNA encodes the necessary information for protein construction, which is then transcribed into RNA before translated into proteins. Regions of the DNA that *directly* encode the construction of proteins are called *genes*. Beyond serving as the blueprint for protein construction, DNA and RNA also encode information that guides regulation of proteins. For instance, they regulate amount of proteins produced (expression level); signals to start or stop production (gene activation or suppression); and signals to modify proteins, affecting their behavior and interaction (protein modification).

2.2 Protein-Protein Interaction (PPI)

Protein, DNA, RNA and other biological molecules do not work in isolation; they cooperate with other proteins to perform a particular biological activity. Two molecules that cooperate to perform a particular function are said to be *interacting*. It is the combination of these molecules and their interactions, and not the molecules alone, that characterize the mechanisms of a biological process. We wish to emphasize that although the rest of the book largely focuses on proteins, the concepts that we will discuss may extend to other molecules. Genes, DNAs, RNAs and other entities are also major drivers of a biological process. Interactions are typically grouped by their molecule types:

- **Protein-protein interactions**—cooperation between proteins to drive biological processes.
- **Gene regulatory interactions**—interplay of genetic information to regulate protein expression level.
- **Metabolic interactions**—cooperation between enzyme proteins to convert a substrate molecule into product molecule through several catalysis reactions.
- RNA-DNA **interactions**—cooperation between RNA-RNA or RNA-DNA interactions plays increasingly critical role in diseases.

In this book, major focus is placed on the class of protein-protein interactions, although most of the concepts covered here apply to other classes of interactions as well.

Protein-protein interactions can be *stable* or *transient* [25]. In *stable* protein-protein interactions, a group of proteins forms permanent protein-protein interactions to perform a biological role. A group of such stably interacting proteins is called a *protein complex*. An example of protein complexes is the `V-ATPase` (Fig. 2.2(a)). Multiple protein subunits combine to form the `V-ATPase` enzyme that transports protons across membranes [24]. In *transient* protein-protein interactions, two proteins associate with each other briefly to perform a biological activity before disassociating. These interactions regulate a significant portion of biological processes. The interactions occur when a region of one protein complements the region of another, forming non-covalent bonds like hydrogen bonds, Van der Waals forces and hydrophobic bondings. A common surface region is the `leucine zipper` [22], a 3D structural motif in proteins with hydrophobic regions that allow two proteins with complementing zipper motifs to "zip" together. Typically, transient interactions only occur under conditions that promote their interaction, for instance the phosphorylation state of the proteins involved, the protein conformation state or their localization. Figure 2.2(b) shows transient interaction between `UBI4` and `PEX12`; physical interaction occurs only during ubiquitination.

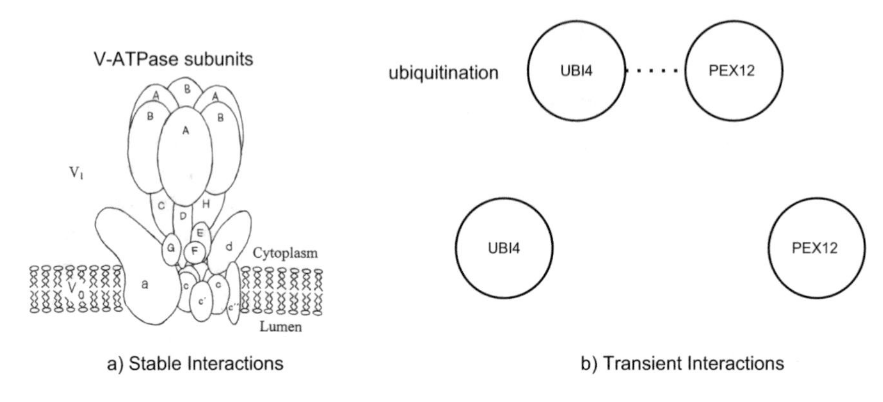

Fig. 2.2 Stable vs. transient interactions

2.3 Methods to Analyze Protein-Protein Interactions

Given the importance of protein-protein interactions in characterizing the mechanisms of a biological process, biologists have developed a range of experimental methods to detect and predict interactions between proteins. We describe several pertinent ones below.

2.3.1 Yeast Two-Hybrid (Y2H)

The yeast-two-hybrid (Y2H) method relies on activating the transcription of a *reporter gene* to detect interaction between two proteins [17]. Reporter genes are typically genes with easily observable phenotype. Figure 2.3 summarizes the concept behind Y2H. In Y2H, biologists engineer the two tested proteins such that when these two proteins interact, transcription of the reporter gene is activated, and thus, if the reporter gene phenotype is sufficiently expressed, one can deduce that the two proteins interact. To this end, Y2H uses two types of protein domains: the DNA-binding domain (BD) and the activation domain (AD). The BD and AD domains must be brought together proximally to bind and form a transcription activator, which then activates reporter gene transcription. Given two proteins, the BD domain is fused to one protein (called the bait) and the AD domain is fused to the remaining protein (called the prey). If these two proteins interact, the two domains are brought together proximally and activates reporter gene transcription. Commonly used reporter genes (and their promoter) include HIS3, URA3 and lacZ. For example, the lacZ reporter gene when activated causes the yeast cell to express β-galactosidase, which can be detected by the formation of blue colored yeast colonies. A strong advantage of this method is its scalability and Y2H can easily be used to screen thousands of proteins for interactions, giving rise to high-throughput experiment technologies.

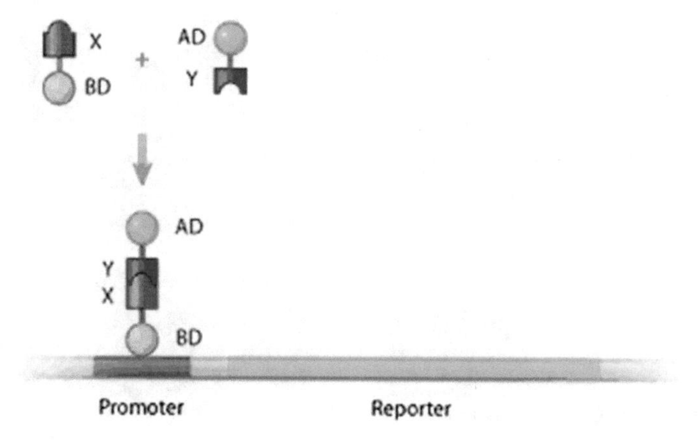

Fig. 2.3 Yeast-two-hybrid to detect protein-protein interactions. (Adopted from *The Science Creative Quarterly*)

2.3.2 *Tandem Affinity Purification (TAP)*

The tandem affinity purification (TAP) method identifies protein-protein interaction by incorporating a TAP tag to the target protein, followed by fishing for other proteins that interact with the tagged protein [28]. Figure 2.4 illustrates the TAP method. The TAP tag comprises two Immunoglobulin G (IgG) binding domains and a Calmodulin-binding peptide (CBP). In TAP, the biologist engineers a fusion protein by fusing the TAP tag to the target protein. Next, the fusion protein, together with any other proteins attached to it, is isolated using beads coated with IgG. The biologist then applies the TEV cleavage enzyme to cleave the TAP tag from fusion, leaving behind the target protein plus the CBP domain bounded to the bead. A second isolation step is then applied using Calmodulin-coated beads. Here, the biologist obtains the final product of target protein, CBP and attached proteins that are interacting with the target protein. Finally, the end products are analyzed via mass spectrometry or SDS-PAGE [31]. The two step purification process minimizes the amount of contaminants obtained.

2.3.3 *Bimolecular Fluorescence Complementation (BIFC)*

Bimolecular fluorescence complementation (BIFC) is another protein-protein interaction screening strategy that relies on a reporter protein [16]. In this method, the reporter protein is fluorescent, allowing it to be easily detected and located using tools such as flow cytometry. A reporter protein, the yellow fluorescent protein (YFP) for instance, is designed as two complementary fragments (YN and YC). Given two candidate proteins, one can separately fuse each fragment to the candidate proteins.

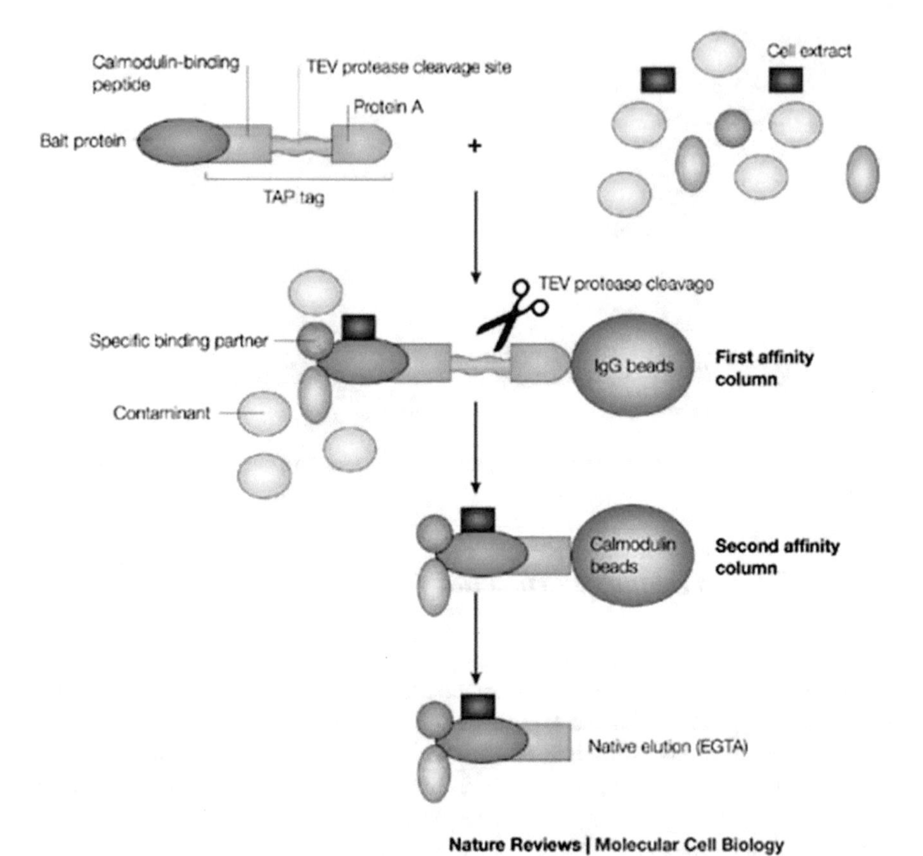

Fig. 2.4 Tandem affinity purification. (Adopted from [15])

When these two proteins interact, the two fragments will be brought to close proximity, encouraging them to re-attach and re-assemble into the YFP reporter protein. The fluorescent reporter protein can then be screened through a variety of techniques including flow cytometry.

2.3.4 Noise in High-Throughput Screening Methods

Rapid high-throughput protein-protein interaction screening methods, however, suffer from significant noise and coverage issues. For instance, the *false negative* rate, defined as the probability of interacting protein detected as negative, could be as high as 70–90% with Y2H data [5]. This imply that there is a significant *coverage* gap (coverage here refers to the ratio between the number of detected interactions and the number of actual interactions in the network). Moreover, high-throughput protein-protein interaction screening methods also suffer from relatively high *false positives* [13], which is defined as the probability of non-interacting protein detected as positive.

2.4 Protein-Protein Interaction Databases

Advancements in protein-protein interaction screening methods have enabled the capability of generating large scale interaction data. Therefore, it is important to catalog and store these datasets to allow rapid and convenient access. We discuss several public databases that catalog key protein-protein interaction datasets. Table 2.1 lists several well known knowledge-bases with significant protein-protein interaction datasets. The STRING database [34] hosts a large collection of predicted and known protein-protein interactions. In addition, the STRING database links key information about the gene that codes for the interactor proteins, including their DNA sequence, biological annotations, co-occurrence, and co-expression data. The Kyoto Encyclopedia of Genes and Genomes (KEGG) database [19] is a resource of manually curated pathway datasets. The KEGG database is especially notable for its large collection of metabolic pathways for bacterial microbes. Important signaling pathways for a variety of organisms are also hosted in the KEGG database. The REACTOME database [18] hosts detailed biological pathways specifically for the human species. As is the KEGG database, pathways in the REACTOME database are manually curated and handcrafted. The IntAct database [21] stores a large

Table 2.1 Selected protein-protein interaction databases

Database	Reference
Human Protein Reference Database (HPRD)	[27]
Biological General Repository for Interaction Datasets (BioGRID)	[32]
Database of Interacting Proteins (DIP)	[35]
Kyoto Encyclopedia of Genes and Genomes (KEGG)	[19]
Biomolecular Interaction Network Database (BIND)	[2]
The MIPS Mammalian Protein-Protein Database	[26]
STRING: functional protein association networks	[34]
REACTOME	[18]
IntAct	[21]
BioCyc	[20]
BioCarta Pathways	[4]
PHOSIDA	[10]
Phospho-ELM	[8]
DOMINE: a database of protein domain interactions	[29]

amount of protein-protein interaction datasets submitted by individual labs. The datasets can range for a several protein-protein interactions per dataset to several hundred thousands of interactions per dataset. The Munich Information Center for Protein Sequences (MIPS) database [26] is noted for its repository of protein complexes. Other significant databases hosting protein-protein interaction datasets are the Human Protein Reference Database (HPRD) [27], Biological General Repository for Interaction Datasets (BioGRID) [32], Database of Interacting Proteins (DIP) [35].

Apart from general protein-protein interaction resources, several web resources host context-specific datasets that focus on a particular biological topic of interest. For example, the PHOSPIDA [10] and Phospho-ELM [8] knowledge-bases contain protein phosphorylation sites information, which can be used to deduce their interacting partners. DOMINE [29] is a database of protein domain-domain interactions. Apart from molecular function specific datasets, disease specific datasets are also abundant. The IntAct database contains a number of disease-related protein-protein interaction datasets that include Alzheimer's, cancer and cerebellar ataxia.

2.5 Annotating the Roles of Proteins and Their Interactions

With the growth of biological literature on the roles of proteins, groups of proteins, as well as their interactions, the need to annotate these information in a structured manner becomes pertinent. The Gene Ontology (GO) [12] is developed as a standard for providing a structured *ontology* describing attributes of genes and gene products (including proteins). An ontology is a set of controlled concepts (GO *terms*) and their relationships that models the domain. In GO, the concepts describe the roles of the genes and their products, while the concept relationships connect the various concepts in GO. For example, the `activation of protein kinase activity` concept can be used to annotate the MAPK protein, giving it that particular function. Now the concept relationships in GO may provide additional inferences to this concept. If suppose GO states that `activation of protein kinase activity` is a type of `regulation of protein phosphorylation`, then one can reason that MAPK protein also has the attribute of `regulation of protein phosphorylation`.

The role of GO as controlled vocabulary also resolves ambiguity in word descriptions. Functional descriptors that describe the role and function of proteins in the literature can be ambiguous, redundant and domain specific [1]. For instance, the gene names `CDC28, Cdc28p or cdc-28` all refer to the same biological entity. With a controlled vocabulary, computation methods can infer functional roles of proteins in a consistent manner.

Gene Ontology Annotation (GOA) database [6] stores associations of genes and proteins to GO terms. GO term annotation can be undertaken either manually or automatically. In manual annotation, a domain expert or curator who is aware of the functional description of the gene or protein annotate that protein with the relevant GO terms. The automatic approach, on the other hand, predicts and infers the GO terms relevant of the protein via a multitude of machine learning techniques including literature mining and graph-based inferencing tools. The Online Mendelian Inheritance in Man (OMIM) database [11] supplies important annotations regarding diseases associated with human proteins.

2.5.1 The Structure of Gene Ontology

The Gene Ontology is modeled as a directed acyclic graph (DAG) and is divided into three major domains: *biological process, cellular component* and *molecular function.* The total number of GO terms in the GO DAG exceeds 30,000.

The `biological process` domain contains GO terms describing the functional processes in cells, tissues, organs and organisms that proteins may take part in. The Gene Ontology defines a biological process as "a recognized series of events or molecular functions" with a defined beginning and end. A biological process GO term may describe the process itself, or it may describe an encompassing process that is made up of subprocesses. For instance, the biological process term `apoptosis` describes cell apoptosis pathways in the cell. Thus, if the `p53` protein is annotated with `apoptosis` GO term, then one can infer that `p53` protein participates in cell apoptosis. The GO term `cell cycle` may describe the cell cycle process which itself is made up of several subprocesses, such as M-phase cell cycle and G-phase cell cycle. In Gene Ontology, a process term may be connected to its parent via `is_a` and `part_of` relationships; the former describes that the process is an instance of the parent process, while the latter describes that the process is only a part of the parent process.

The `cellular component` domain contains GO terms describing the components of the cell and its extracellular environment. Cellular components may be anatomical structures or macromolecular complexes. In GO, a protein annotated with a cellular component GO term is said to be *located in* or is a *subcomponent of* the component described by the term. For example, the GO term `mitochrondrial ribosome` describes the mitochondrial ribosomal components. Proteins like ribosomal protein `L41` may be annotated with such GO terms.

Finally, the `molecular process` contains GO terms pertaining to an elemental activity of a protein. Activities here include any function performed by proteins like catalysis, binding, phosphorylation, and other enzymatic roles. For example, the GO term `phosphorylation` describes the molecular activity that a protein may perform, which in this case is phosphorylation activity. A protein may be annotated with multiple activities. This is because proteins itself may participate in multiple functions. Protein kinases like `PKC` are known to have such capabilities and could be annotated with these terms.

Figure 2.5 depicts a part of the GO DAG. Formally, the Gene Ontology for each domain is modeled as a directed acyclic graph $D = (V_{go}, E_{go})$ where V_{go} denotes

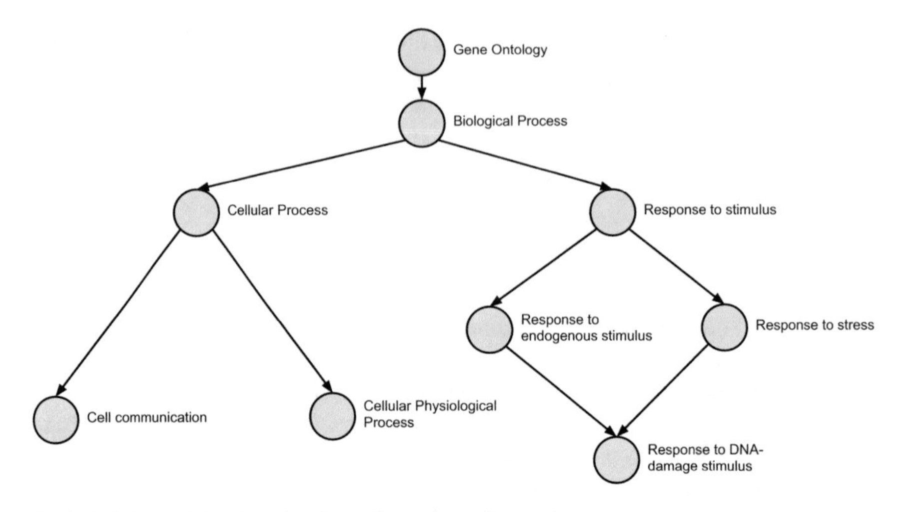

Fig. 2.5 Subset of the Gene Ontology directed acyclic graph

the set of GO terms and E_{go}—the set of pair relationships between GO terms in V_{go}—denotes the set of GO term relationships. Here, an edge $(v_1, v_2) \in E$ represents a parent-child connection between two GO terms $v_1 \in V_{go}$ and $v_2 \in V_{go}$. The ordered set $\Delta = \langle \Delta_1, \Delta_2, \ldots, \Delta_d \rangle$ is a topological sort of D. Each Δ_i represents a single GO term. We assume that a protein node $v \in V_i$ is annotated with a set of GO terms $D_v \subset \Delta$. The indicator function of terms annotated in node v is given by $I_{\{x \in D_v\}} : \Delta \to \{0, 1\}$ such that $I_{\{x \in D_v\}} = 1$ if $x \in D_v$ and 0 if otherwise.

The root node absorbs all GO terms of its descendants, i.e., each descendant GO term is a or is part of the root node. The root nodes of biological process, cellular component and molecular function domains are `biological process`, `cellular component` and `molecular function`, respectively. As the GO DAG branches from the root node, the specificity of the functional description increases. Thus, one can utilize GO DAG and its associated annotations to group proteins by their function or parts in a hierarchical manner. For example, in Fig. 2.5, if proteins MAPK, MAPKK, and MAPKKK are annotated with the `intracellular signaling` process term, then these proteins are also part of the `signal transduction`, `cell communication`, `cellular process` and `biological process`.

Gene Ontology and its annotations has been applied to a large number of bioinformatics approaches [14]. A pertinent usage is in gene expression analysis studies [3]. Typically, groups of genes which are either significantly up-regulated or de-regulated are identified using techniques such as gene clustering and enrichment analysis [33]. Then, the GO annotations are utilized to identify over-expressed functional roles of these groups of proteins. An example of algorithms of such nature is the `MAPPFinder` [9], which looks for genes that are significantly deregulated using the GO annotations.

2.6 Summary

This chapter can be summarized as follows:

- Proteins, DNAs, RNAs and other biological molecules work in tandem to regulate biological processes. Cooperating molecules that perform a particular function are said to be interacting, and their interactions can be either transient or stable.
- A range of experimental methods have been developed to detect and predict interactions between proteins in a high-throughput manner. Among them are `Y2H`, `TAP` and `BIFC`.
- Advancement in protein-protein interaction screening methods has led to large scale interaction datasets. Several public databases serve as important repositories of such datasets, including `STRING`, `KEGG` and `REACTOME`.
- The Gene Ontology (GO) is developed as a standard for providing a structured ontology describing attributes of genes and gene products (including proteins). Gene Ontology Annotation (GOA) database stores associations of genes and proteins to GO terms. GO term annotations are useful as functional descriptions of a gene or protein.

References

1. M. Ashburner, C.A. Ball, J.A. Blake, D. Botstein, H. Butler, J.M. Cherry, A.P. Davis, K. Dolinski, S.S. Dwight, J.T. Eppig, M.A. Harris, D.P. Hill, L. Issel-Tarver, A. Kasarskis, S. Lewis, J.C. Matese, J.E. Richardson, M. Ringwald, G.M. Rubin, G. Sherlock, Gene Ontology: tool for the unification of biology. the Gene Ontology consortium. Nat. Genet. **25**, 25–29 (2000)
2. G.D. Bader, I. Donaldson, C. Wolting, B.F. Ouellette, T. Pawson, C.W. Hogue, BIND-the biomolecular interaction network database. Nucl. Acids Res. **29**, 242–245 (2001)
3. T. Beissbarth, T.P. Speed, GOstat: find statistically overrepresented Gene Ontologies within a group of genes. Bioinformatics (Oxford, England) **20**, 1464–1465 (2004)
4. Biocarta, BioCarta
5. P. Braun, M. Tasan, M. Dreze, M. Barrios-Rodiles, I. Lemmens, H. Yu, J.M. Sahalie, R.R. Murray, L. Roncari, A.-S. de Smet, K. Venkatesan, J.-F. Rual, J. Vandenhaute, M.E. Cusick, T. Pawson, D.E. Hill, J. Tavernier, J.L. Wrana, F.P. Roth, M. Vidal, An experimentally derived confidence score for binary protein-protein interactions. Nat. Methods **6**, 91–97 (2009)
6. E. Camon, M. Magrane, D. Barrell, V. Lee, E. Dimmer, J. Maslen, D. Binns, N. Harte, R. Lopez, R. Apweiler, The Gene Ontology annotation (GOA) database: sharing knowledge in uniprot with Gene Ontology. Nucl. Acids Res. **32**, D262–D266 (2004)
7. F. Crick, Central dogma of molecular biology. Nature **227**, 561–563 (1970)
8. H. Dinkel, C. Chica, A. Via, C.M. Gould, L.J. Jensen, T.J. Gibson, F. Diella, Phospho.ELM: a database of phosphorylation sites-update 2011. Nucl. Acids Res. **39**, D261–D267 (2011)

9. S.W. Doniger, N. Salomonis, K.D. Dahlquist, K. Vranizan, S.C. Lawlor, B.R. Conklin, MAPPFinder: using Gene Ontology and GenMAPP to create a global gene-expression profile from microarray data. Gen. Biol. **4**, R7–R7 (2003)
10. F. Gnad, S. Ren, J. Cox, J.V. Olsen, B. Macek, M. Oroshi, M. Mann, PHOSIDA (phosphorylation site database): management, structural and evolutionary investigation, and prediction of phosphosites. Gen. Biol. **8**, R250–D250 (2007)
11. A. Hamosh, A.F. Scott, J.S. Amberger, C.A. Bocchini, V.A. McKusick, Online mendelian inheritance in man (OMIM), a knowledgebase of human genes and genetic disorders. Nucl. Acids Res. **33**, D514–D517 (2005)
12. M.A. Harris, J. Clark, A. Ireland, J. Lomax, M. Ashburner, R. Foulger, K. Eilbeck, S. Lewis, B. Marshall, C. Mungall, J. Richter, G.M. Rubin, J.A. Blake, C. Bult, M. Dolan, H. Drabkin, J.T. Eppig, D.P. Hill, L. Ni, M. Ringwald, R. Balakrishnan, J.M. Cherry, K.R. Christie, M.C. Costanzo, S.S. Dwight, S. Engel, D.G. Fisk, J.E. Hirschman, E.L. Hong, R.S. Nash, A. Sethuraman, C.L. Theesfeld, D. Botstein, K. Dolinski, B. Feierbach, T. Berardini, S. Mundodi, S.Y. Rhee, R. Apweiler, D. Barrell, E. Camon, E. Dimmer, V. Lee, R. Chisholm, P. Gaudet, W. Kibbe, R. Kishore, E.M. Schwarz, P. Sternberg, M. Gwinn, L. Hannick, J. Wortman, M. Berriman, V. Wood, N. de la Cruz, P. Tonellato, P. Jaiswal, T. Seigfried, R. White, The Gene Ontology (GO) database and informatics resource. Nucl. Acids Res. **32**, D258–D261 (2004)
13. H. Huang, J.S. Bader, Precision and recall estimates for two-hybrid screens. Bioinformatics (Oxford, England) **25**, 372–378 (2009)
14. D.W. Huang, B.T. Sherman, R.A. Lempicki, Systematic and integrative analysis of large gene lists using DAVID bioinformatics resources. Nat. Protoc. **4**, 44–57 (2009)
15. L.A. Huber, Is proteomics heading in the wrong direction? Nat. Rev. Mol. Cell Biol. **4**, 74–80 (2003)
16. C.-D. Hu, Y. Chinenov, T.K. Kerppola, Visualization of interactions among bZIP and Rel family proteins in living cells using bimolecular fluorescence complementation. Mol. Cell **9**, 789–798 (2002)
17. T. Ito, T. Chiba, R. Ozawa, M. Yoshida, M. Hattori, Y. Sakaki, A comprehensive two-hybrid analysis to explore the yeast protein interactome. in *Proceedings of the National Academy of Sciences of the United States of America*, vol. 98 (2001), pp. 4569–4574
18. G. Joshi-Tope, M. Gillespie, I. Vastrik, P. D'Eustachio, E. Schmidt, B. de Bono, B. Jassal, G.R. Gopinath, G.R. Wu, L. Matthews, S. Lewis, E. Birney, L. Stein, Reactome: a knowledgebase of biological pathways. Nucl. Acids Res. **33**, D428–D432 (2005)
19. M. Kanehisa, S. Goto, KEGG: kyoto encyclopedia of genes and genomes. Nucl. Acids Res. **28**, 27–30 (2000)
20. P.D. Karp, C.A. Ouzounis, C. Moore-Kochlacs, L. Goldovsky, P. Kaipa, D. Ahrén, S. Tsoka, N. Darzentas, V. Kunin, N. López-Bigas, Expansion of the BioCyc collection of pathway/genome databases to 160 genomes. Nucl. Acids Res. **33**, 6083–6089 (2005)
21. S. Kerrien, Y. Alam-Faruque, B. Aranda, I. Bancarz, A. Bridge, C. Derow, E. Dimmer, M. Feuermann, A. Friedrichsen, R. Huntley, C. Kohler, J. Khadake, C. Leroy, A. Liban, C. Lieftink, L. Montecchi-Palazzi, S. Orchard, J. Risse, K. Robbe, B. Roechert, D. Thorneycroft, Y. Zhang, R. Apweiler, H. Hermjakob, IntAct-open source resource for molecular interaction data. Nucl. Acids Res. **35**, D561–D565 (2007)
22. W.H. Landschulz, P.F. Johnson, S.L. McKnight, The leucine zipper: a hypothetical structure common to a new class of DNA binding proteins. Science (New York, N.Y.) **240**, 1759–1764 (1988)
23. J. Lodish, H.F. Baltimore, D. Berk, A. Darnell, *Molecular Cell Biology* (WH Freeman, New York, 1995)
24. V. Marshansky, M. Futai, The V-type H+-ATPase in vesicular trafficking: targeting, regulation and function. Curr. Opin. Cell Biol. **20**, 415–426 (2008)
25. I.M.A. Nooren, J.M. Thornton, Diversity of protein-protein interactions. EMBO J. **22**, 3486–3492 (2003)

26. P. Pagel, S. Kovac, M. Oesterheld, B. Brauner, I. Dunger-Kaltenbach, G. Frishman, C. Montrone, P. Mark, V. Stümpflen, H.-W. Mewes, A. Ruepp, D. Frishman, The MIPS mammalian protein-protein interaction database. Bioinformatics (Oxford, England) **21**, 832–834 (2005)
27. S. Peri, J.D. Navarro, R. Amanchy, T.Z. Kristiansen, C.K. Jonnalagadda, V. Surendranath, V. Niranjan, B. Muthusamy, T.K.B. Gandhi, M. Gronborg, N. Ibarrola, N. Deshpande, K. Shanker, H.N. Shivashankar, B.P. Rashmi, M.A. Ramya, Z. Zhao, K.N. Chandrika, N. Padma, H.C. Harsha, A.J. Yatish, M.P. Kavitha, M. Menezes, D.R. Choudhury, S. Suresh, N. Ghosh, R. Saravana, S. Chandran, S. Krishna, M. Joy, S.K. Anand, V. Madavan, A. Joseph, G.W. Wong, W.P. Schiemann, S.N. Constantinescu, L. Huang, R. Khosravi-Far, H. Steen, M. Tewari, S. Ghaffari, G.C. Blobe, C.V. Dang, J.G.N. Garcia, J. Pevsner, O.N. Jensen, P. Roepstorff, K.S. Deshpande, A.M. Chinnaiyan, A. Hamosh, A. Chakravarti, A. Pandey, Development of human protein reference database as an initial platform for approaching systems biology in humans. Gen. Res. **13**, 2363–2371 (2003)
28. O. Puig, F. Caspary, G. Rigaut, B. Rutz, E. Bouveret, E. Bragado-Nilsson, M. Wilm, B. Séraphin, The tandem affinity purification (TAP) method: a general procedure of protein complex purification. Methods (San Diego, Calif.) **24**, 218–229 (2001)
29. B. Raghavachari, A. Tasneem, T.M. Przytycka, R. Jothi, DOMINE: a database of protein domain interactions. Nucl. Acids Res. **36**, D656–D661 (2008)
30. G. Scatchard, The attractions of proteins for small molecules and ions. Ann. New York Acad. Sci. **51**, 660–672 (1949)
31. H. Schägger, Tricine-SDS-PAGE. Nat. Protoc. **1**, 16–22 (2006)
32. C. Stark, B.-J. Breitkreutz, T. Reguly, L. Boucher, A. Breitkreutz, M. Tyers, BioGRID: a general repository for interaction datasets. Nucl. Acids Res. **34**, D535–D539 (2006)
33. A. Subramanian, P. Tamayo, V.K. Mootha, S. Mukherjee, B.L. Ebert, M.A. Gillette, A. Paulovich, S.L. Pomeroy, T.R. Golub, E.S. Lander, J.P. Mesirov, Gene set enrichment analysis: a knowledge-based approach for interpreting genome-wide expression profiles. in *Proceedings of the National Academy of Sciences of the United States of America*, vol. 102 (2005), pp. 15545–15550
34. D. Szklarczyk, A. Franceschini, M. Kuhn, M. Simonovic, A. Roth, P. Minguez, T. Doerks, M. Stark, J. Muller, P. Bork, L.J. Jensen, C. von Mering, The STRING database in 2011: functional interaction networks of proteins, globally integrated and scored. Nucl. Acids Res. **39**, D561–D568 (2011)
35. I. Xenarios, D.W. Rice, L. Salwinski, M.K. Baron, E.M. Marcotte, D. Eisenberg, DIP: the database of interacting proteins. Nucl. Acids Res. **28**, 289–291 (2000)
36. T.S. Young, P.G. Schultz, Beyond the canonical 20 amino acids: expanding the genetic lexicon. J. Biol. Chem. **285**, 11039–11044 (2010)

Chapter 3
Clustering PPI Networks

Due to the availability of large-scale PPI networks, since the last decade significant research efforts have been invested in analyzing these networks in order to comprehend cellular organization and functioning [1]. Among myriads of such efforts, *network clustering* (or graph clustering) is arguably one of the most popular approaches for analyzing the topological and functional properties of a PPI network. Specifically, the goal here is to identify *clusters*, subgraphs of the PPI network that exhibit significant *clustering properties*. These clusters enable us to uncover the following modules:

- *Functional modules*: These are collection of proteins of similar or related functional properties in the same network neighborhood. That is, a functional module represent a collection of molecular interactions that work together to achieve a particular functional objective in a biological process. These modules may be pathways, protein complexes or other biological processes. An example of such pathway is the MAPK signaling, a collection of interacting proteins that act as messengers to amplify and distribute signals from stimuli to intended destinations.
- *Topological modules*: These modules represent locally dense neighborhood in a PPI network. Particularly, vertices in a topological module have a higher tendency to connect to other vertices within the same local neighborhood than to vertices outside it. Note that these modules are oblivious to the function of individual proteins.

Hence, PPI network clustering can be considered as summarizing a PPI network with respect to its topological and functional modules. Furthermore, PPI network clustering may also enable us to infer function of a protein by assigning to it the function of another protein which belongs to the same cluster. Figure 3.1 illustrates a clustering of a PPI network, showing the `RSC complex` and `SWI/SNF complex` proteins grouped into distinct clusters. The shape of a protein node indicates an assignment of the protein into either `RSC complex` cluster (circle nodes)

S.S. Bhowmick and B.-S. Seah, *Summarizing Biological Networks*,
Computational Biology 24, DOI 10.1007/978-3-319-54621-6_3

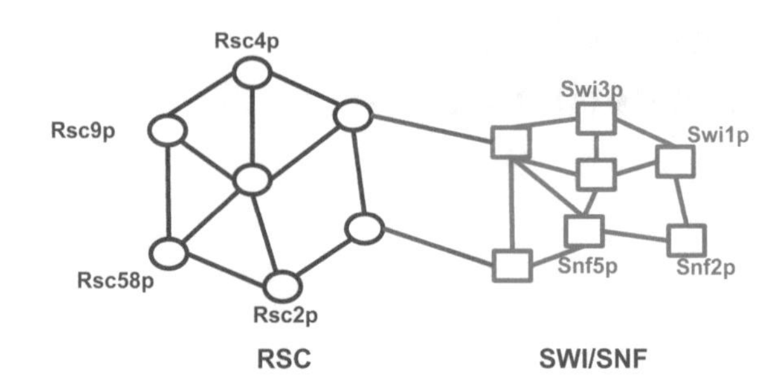

Fig. 3.1 An example of PPI network clustering

or `SWI/SNF complex` cluster (square nodes). Note that the number of protein-protein interaction connections within a cluster is significantly higher than those which are between clusters. Despite the computational complexity of the problem of identifying such modules, a wide spectrum of PPI network clustering algorithms with various characteristics have emerged since the last decade. In this chapter, we give a comprehensive review of these [2] techniques.

The rest of the chapter is organized as follows. In the next section, we formally introduce the PPI network clustering problem and associated concepts. In Sect. 3.2, we present an overview of key PPI clustering approaches proposed in the literature, highlighting their respective strengths and weaknesses. We introduce some of the key measures used in the literature to validate predicted clusters by these techniques in Sect. 3.3. We provide a comparative summary of these techniques in Sect. 3.4. The last section concludes this chapter. Due to the cornucopia of techniques proposed in the literature for PPI network clustering, we do not focus on exhaustive performance comparison of various clustering techniques in this chapter. The reader may refer [3] to get a glimpse of such performance study.

3.1 PPI Network Clustering Problem

In this section, we first formally introduce the PPI *network clustering* problem and its associated terminology that we shall be using in the sequel. Next, we articulate the challenges associated with clustering PPI networks and why generic network clustering algorithms that are not specifically designed for such networks fail to provide effective solutions to this problem. Then, we briefly identify the representative clustering measures used in the literature to cluster PPI networks. Lastly, we present a brief overview and classification of different PPI clustering techniques proposed in the literature. In the next section, we shall review these techniques in detail.

3.1.1 Problem Definition

A protein-protein interaction (PPI) network can be modeled as an undirected graph $G = (V, E, \omega)$ that contains a set of vertices (or nodes) V representing proteins and a set of edges E representing interactions. The function ω assigns each interaction $e \in E$ a weight that represents its interaction strength or confidence. Formally, a *clustering* of G aims to partition V into a set of clusters $\mathbb{C} = \{C_1, C_2, \ldots, C_d\}$ that maximizes *clustering property* objective function $f : \mathbb{C} \rightarrow \mathbb{R}$, i.e., find any $\mathbb{C}$ in $argmax_{\mathbb{C}} f(\mathbb{C}) = \{\mathbb{C} | \forall \mathbb{C}' : f(\mathbb{C}') \leq f(\mathbb{C})\}$. Typically, $f(\cdot)$ rewards clusters that exhibit many within-cluster edges and few between-cluster edges. Quantitatively, it can be defined as follows [4]. Let $d_{in}(v_i)$ and $d_{out}(v_i)$ be the number of edges connecting a vertex $v_i \in C_k$ to other vertices in a cluster C_k and the number of edges connecting v_i to vertices in G that are not in C_k, respectively. Then, C_k is a *strong cluster* if $\forall v_i \in C_k$:

$$d_{in}(v_i) > d_{out}(v_i) \tag{3.1}$$

On the other hand, C_k is a *weak cluster* if:

$$\sum_{v_i \in C_k} d_{in}(v_i) > \sum_{v_i \in C_k} d_{out}(v_i) \tag{3.2}$$

Recall that pathways and processes in a biological system do not work in isolation; instead they work in tandem to coordinate the functionalities of the cell. Moreover, it is possible to organize these processes into even higher-order processes, forming a hierarchy of biological processes [5]. Hence the key objective of PPI network clustering is to analyze the topological and functional properties of a PPI network to identify and predict potential topological and functional modules represented as clusters.

A clustering is *disjoint* or *non-overlapping* if the vertices in $\mathbb{C}$ forms partition of V, i.e., $\forall C_i, C_j \in \mathbb{C}$, $V_i \cap V_j = \emptyset$ and $\bigcup_{k \in [1, |\mathbb{C}|]} V_k = V$. On the other hand, a clustering has *overlapping* clusters if $\exists C_i, C_j \in \mathbb{C}$, $V_i \cap V_j \neq \emptyset$. Clustering algorithms may be characterized by whether the clustering obtained are disjoint or overlapping. Specifically, in overlapping clusters, different modules are allowed to share the same vertices in the network. Non-overlapping clustering, on the other hand, constructs modules that share no vertices.

3.1.2 Challenges

The unique characteristics of PPI networks demand that clustering techniques designed for these networks should consider the following distinguishable features of the network and clusters.

- *Overlapping nature of clusters.* In recent years the idea of "one gene-one protein-one function" has been superseded by the knowledge that many proteins have multiple functions. One category of such proteins is called "moonlighting" proteins which includes diverse set of enzymes, chaperones, transcription factors, and proteins with many other types of functions (e.g., `Ubp6`). Consequently, clusters in a PPI network may overlap where a protein is involved in multiple functional modules. This necessitates that PPI clustering techniques should generate overlapping clusters.
- *Edge weights of interactions.* Recall that PPI networks are incomplete and noisy due to the limitations in the experimental procedures. Hence, edges of a PPI network may be associated with weights to model the uncertainty associated with the interactions. A clustering algorithm should consider this noisiness during cluster detection.
- *Attributed* PPI *networks.* Recall that with the growth of biological literature on the roles of proteins, groups of proteins, as well as their interactions, increasingly nodes in a PPI network are annotated with attributes (such as Gene Ontology (GO)) to encode information such as functions, localization, and biological processes that they are involved with. The richness provided by this (partially) attributed PPI networks introduces additional challenges to their clustering (for example, the high dimensionality of protein attributes), but at the same time, it opens the door for opportunities to yield novel findings. Hence, clustering techniques should leverage such annotations whenever available to create superior quality clusters.
- *Dense and sparse clusters.* As remarked earlier, although topological modules may be dense but functional modules may not be. Specifically, pathways may be sparsely connected to perform certain functions. Hence, a PPI clustering technique needs to consider both dense and sparse clusters for identifying superior quality results.
- *Full coverage of the* PPI *network.* It is important for a clustering technique to cover all the nodes of a PPI network as clusters can be both dense and sparse. This will ensure that important functional modules or protein complexes are not missed during the clustering process.
- *Scalability.* With the advent of high-throughput experimental techniques, information on thousands of PPI are being generated. It is estimated that the complete set of protein interactions for humans contains 650,000 interactions [6]. Hence scalable tools are necessary to cluster such large PPI networks.

There has been significant effort by the data mining community to address the generic network clustering or community detection problem. For instance, spectral techniques [7–9] discover dense network modules and bipartite structures by performing recursive bisection and multiway partitioning based on the Fiedler vector of the graph Laplacian. Brandes et al. [10] have proposed minimum cuts or maximum flows-based technique to detect clusters (community). There are also several techniques that are density-based [11, 12] or leverage hierarchical agglomerative clustering principle to detect community structures [4, 13–15]. Graph partitioning-based approaches [16, 17] aim to partition the network into equal or unequal size partitions using techniques such as weighted k-means. More recently, the data mining community has focused on clustering attributed networks [18]. Unfortunately, these generic network clustering techniques cannot be adopted easily to cluster PPI networks as they do not address one or more of the aforementioned issues. For instance, several of these techniques do not generate overlapping clusters, or only identify dense clusters, or do not consider edge weights and node attributes. This has led to a large body of work focusing on techniques specifically designed to cluster PPI networks.

3.1.3 Representative Clustering Measures

In this section, we briefly describe key measures that are used in the literature to determine cluster memberships in PPI networks.

Clustering Coefficient. Informally, *clustering coefficient* represents the interconnectivity of a vertex v's neighbors. Let t_v be the number of triangles that involve v and k_v be the degree of v. The clustering coefficient of v is defined as follows:

$$CC(v) = \frac{2t_v}{k_v \times (k_v - 1)} \tag{3.3}$$

Density. The *density* of a subgraph or cluster $C = (V_C, E_C)$ is defined as the ratio of the number of edges in C over the maximum number of possible edges in C:

$$Density(C) = \frac{2|E_C|}{|V_C| \times (|V_C| - 1)} \tag{3.4}$$

As we shall see later, many clustering algorithms utilize cluster density to identify topological modules—subgraphs with density that exceed a specific *density threshold*. The density of a cluster can be weighted. In that case, the *weighted density* of C is given by:

$$Weighted\ Density(C) = \frac{2\sum_{e \in E_C} \omega(e)}{|V_C| \times (|V_C| - 1)} \tag{3.5}$$

Cluster Cohesiveness. *Cluster cohesiveness* [19] measures how likely it is for a group of proteins to form a protein complex. It is defined formally as follows. Let V be a group of proteins. Let $w^{in}(V)$ be the aggregated weight of edges induced by V. Since V is connected to the rest of the PPI network, let $w^{bound}(V)$ be the aggregated weight of edges E_w where for each $(u, v) \in E_w$, $u \in V$ and $v \notin V$. Then, the *cluster cohesiveness* of V is given by:

$$f(V) = \frac{w^{in}(V)}{w^{in}(V) + w^{bound}(V) + p|V|} \tag{3.6}$$

The term $p|V|$ in the above equation is a penalty term used to capture the uncertainty associated with the PPI network as certain interactions may yet to be discovered due to the limitations in the experimental procedures.

Conductance. Informally, *conductance* [17] refers to clusters that have many edges within it and few edges going out of the clusters. The *conductance* of a set S is given by:

$$\Upsilon(S) = \frac{\sigma(S)}{min(vol(S), 2m - vol(S))} \tag{3.7}$$

where $\sigma(s) = |\{(u, v)|u \in S, v \notin S\}|$; $vol(S) = \sum_{v \in S} k_v$ and $m = |E|$. The volume of S, $vol(S)$, is also written as $\mu(S)$. The *conductance* of G is then given by:

$$\Upsilon_G = min_{S \subset V} \Upsilon(S) \tag{3.8}$$

Log Odd Score. This measure quantifies the association between a pair of proteins that are indirectly related through *shared neighbors*—two proteins are deemed highly connected if they share most of their neighbors. Specifically, the *log odd score* r_{uv} of a protein pair (u, v) is given as follows [20]:

$$r_{uv} = ln \frac{P(s_{uv}|\hat{\lambda})}{P(s_{uv}|\bar{\lambda})} \tag{3.9}$$

where s_{uv} is the observed number of shared neighbors between the protein pair, $\bar{\lambda}$ and $\hat{\lambda}$ are the Poisson parameters of s_{uv} under the null and alternative hypothesis, respectively. Note that the null hypothesis in this case is that the number of shared neighbors between the protein pair from a random network whereas the alternative hypothesis is the number of such pairs is greater than expected from a random network.

Czeknowski-Dice Distance. It is a neighborhood-based similarity measure for clustering PPI networks and is based on the intuition that vertices that have common or shared neighbors may have some degree of similarity even if they do not have any direct interaction. Formally, the Czeknowski-Dice distance [21, 22] of proteins v_i and v_j is defined as

$$S_n(v_i, v_j) = \frac{|Int(i) \Delta Int(j)|}{|Int(i) \cup Int(j)| + |Int(i) \cap Int(j)|} \tag{3.10}$$

where $Int(i)$ is the adjacent list of protein i and Δ is the symmetric difference between the sets.

Edge Betweenness. Betweenness centrality is widely recognized as an important global metric for analyzing topological characteristics of many real-world networks. Specifically, in the context of PPI networks, *edge betweenness* of an edge is the number of shortest paths that pass through it normalized by the maximum number of shortest paths through an edge. Formally, edge betweenness of (v_i, v_j) can be defined as follows where SP_{ij} and SP_{max} denote the number of shortest path through it and maximum number of shortest paths, respectively [21]:

$$S_{eb}(v_i, v_j) = 1 - \frac{SP_{ij}}{SP_{max}} \tag{3.11}$$

Variational Information Distance. This measure leverages annotations associated with proteins to compute distance between clusters. Let C be a clustering on the network G and D be a clustering of V by their MIPS annotation of the proteins. The *variational information distance* [23] between two clusterings is defined as:

$$VI(C, D) = H(C) + H(D) - 2I(C, D) \tag{3.12}$$

where $H(C)$ and $H(D)$ are the entropies of C and D, respectively; and $I(C, D)$ is the mutual information between C and D. Intuitively, the entropies measure the amount of uncertainty or information in each clustering, while $I(C, D)$ measures how much information is shared among C and D. Thus, $VI(C, D)$ measures how much uncertainties are encoded in C given that D is known.

Modularity. Given a PPI network G and a clustering C, the *modularity* of C [24] is defined as:

$$q(G, C) = \sum_{u,v \in V} \frac{I_{uv} - d(u) \times d(v)/2|E|}{1 - x_{uv}} \tag{3.13}$$

where $I_{uv} = 1$ if $(u, v) \in E$ or 0 otherwise; $d(.)$ is the node degree; and $x_{uv} = 1$ if u and v belong to the same cluster and 0 otherwise. Note that the above objective function rewards groups of proteins with strong co-connections when they are placed in the same cluster.

3.1.4 Overview

Table 3.1 presents an overview of techniques to address the PPI network clustering problem. Specifically, we classify these efforts into the following categories to facilitate our discussion.

- Heuristic-based Algorithms
- Flow-based Algorithms
- Complete Enumeration Algorithms
- Random Walks and Message Passing Algorithms
- Graph-cut and Hierarchical Clustering Algorithms
- Multiple Clustering-based Algorithms
- Genomic Data-driven Clustering Algorithms

Table 3.1 Overview of PPI network clustering techniques

Algorithms		Description	Datasets
Heuristic-based Algorithms			
MCODE	[25]	Local neighborhood density based	[26], MIPS, YPD [27], Y2H exp. [28–32]
DPClus	[33]	Local density and periphery based	DIP [34], MIPS
IPCA	[35]	Local density and distance based	MIPS
RNSC	[36]	Local search cost based	Yeast [37], Fruitfly [38], *C. elegans* [39], MIPS
Pei et al.	[40]	Subgraph refinement based	[26], DIP, Y2H experiments [28–32]
ClusterONE	[19]	Cluster cohesiveness based	[41–43], BioGRID
SPICi	[44]	Weighted density based	BioGRID, STRING [45], Bayesian network [46]
Flow-based Algorithms			
TRIBE-MCL	[47]	Flow simulation	InterPro, SwissProt
MLR-MCL	[48]	Flow simulation	DIP, BioGRID, iRefIndex [49]
SR-MCL	[50]	Iterative flow simulation	DIP, BioGRID, WI- PHI [51]
Pereira-Leal et al.	[52]	Flow simulation on line graph	DIP
Cho et al.	[53]	Flow simulation using informative proteins	DIP, MIPS, Stanford Microarray Database
Cho et al.	[54]	Flow pattern mining	MIPS, GO
IQ-flow	[55]	Integrates functional flow and quantum-behaved Particle Swarm Optimization (QPSO)	MIPS
Complete enumeration algorithms			
SPC and MC	[56]	Maximal clique based	MIPS
CFinder	[57]	k-clique based	DIP
Zhang et al.	[58]	CPM based	MIPS
Cui et al.	[59]	Quasi-clique based	MIPS, [32], [29], FunCat [60]
CMC	[61]	Maximal clique based	[26], [32], [62], [42], [29], MIPS, [63]
DME	[64]	Module density based	DIP, MPact [65], BIND, HPRD, MINT, IntAct [66], [26, 42]

(continued)

Table 3.1 (continued)

Algorithms		Description	Datasets
Random walks and message passing algorithms			
Affinity Propagation (AP)	[67]	Based on availability and responsibility	yeast network, [43]
Nibble	[17]	Random walk and conductance based	BioGRID
RRW	[68]	Random walk with restart based	MIPS, WI- PHI [51]
Graph-cut and hierarchical clustering algorithms			
Chen et al.	[69]	Edge betweenness based	[29, 32, 37, 62], MIPS
MoNet	[5]	Hierarchical module based	DIP, Saccharomyces Genome Database (SGD) [70], GO
Tree-Snipping	[71]	Annotation-driven hierarchical clustering	GO
VI-Cut	[23]	Variational information based	IntAct, MIPS
SCAN	[72]	Structural clustering based on common neighbors	SGD, GO
NeMo	[20]	Based on shared neighborhood	MIMI [73]
Multiple clustering-based algorithms			
Asur et al.	[21]	Ensemble clustering based	DIP
Greene et al.	[74]	Ensemble clustering based	[43], MIPS
MOD-ILP	[24]	Integer linear programming based	Only signaling network data is used
Genomic data-driven clustering algorithms			
Segal et al.	[75]	Integrates gene expressions	Gene expression data [76, 77], DIP, GO
Lu et al.	[78]	Hierarchical clustering with expression data	Yeast network, subcellular location data [79], expression profile [80], MIPS
Maraziotis et al.	[81]	Density based	[37, 82], yeast expression [77], MIPS
Zheng et al.	[83]	Bayesian network based	genomic features [84], experimental interaction [29, 32, 41, 62], mRNA expression [85, 86], MIPS, SGD, GO
CEZANNE	[87]	Probabilistic model based on minimum cut	Yeast expression profile [88], [43], GO, MIPS, SGD
Shi et al.	[89]	Neural network based	DIPS, BioGRID, MIPS, GO

Heuristic-based algorithms utilize a greedy heuristic to identify clusters. In *Flow-based* algorithms, clustering is achieved by a series of flow "expansions" and "contractions" to identify clusters with high intra-cluster flows and weak inter-cluster flows. *Complete Enumeration* algorithms enumerate all possible subgraphs with density exceeding a specified threshold. *Random Walk-based* methods model the graph clustering problem as identifying the stationary distribution of a random walk model. On the other hand, *Graph-cut and Hierarchical Clustering* algorithms utilize graph theoretic properties to identify clusters. The *Multiple Clustering-based* algorithms generate a set of clustering instead of a single clustering and combine or investigate them to create the final clustering. Lastly, *Genomic Data-driven Clustering* algorithms integrate genomic and PPI data to address the problem of noise in PPI networks.

Table 3.1 also presents the datasets used by these work. Observe that a wide variety of PPI datasets are used by these algorithms to experimentally evaluate their clustering quality. Among these, DIP, BioGRID, and MIPS are widely used by several techniques across these categories. Details related to these popular datasets are given in [3]. On the other hand, several datasets (e.g., MPact [65], WI- PHI [51]) are not so popular among various techniques.

3.2 PPI Network Clustering Techniques

In this section we describe representative PPI clustering techniques of the aforementioned categories and discuss the features that distinguish them.

3.2.1 Heuristic-Based Algorithms

Heuristic-based algorithms find dense network regions by searching heuristically for potential cluster regions using an iterative greedy *seed and extend* strategy. One of the seminal efforts is MCODE [25], which identifies densely connected clusters based on a seed and extend heuristic. In this approach, a weighing scheme is introduced that searches for dense local neighborhood regions. Given a PPI network $G = (V, E)$, the MCODE algorithm consists of three key phases as follows.

Phase 1: Vertex weighting. Let the 1-neighborhood of a protein $u \in V$ be the subgraph $N(u) = (V_u, E_u)$ induced by the vertex u and its immediate neighborhood. For each $v \in V$, MCODE identifies the highest $k - core$ number of the 1-neighborhood of v. A *k-core* is a graph $G_k = (V_k, E_k)$ such that for all $v_k \in V_k$, $d(v_k) \leq k$ where $d(v_k)$ is the degree of v_k. The highest *k-core* number of the 1-neighborhood of v is then defined as the largest number k such that the subgraph is *k-core*. Furthermore, MCODE determines for the 1-neighborhood of v its *density*, given by:

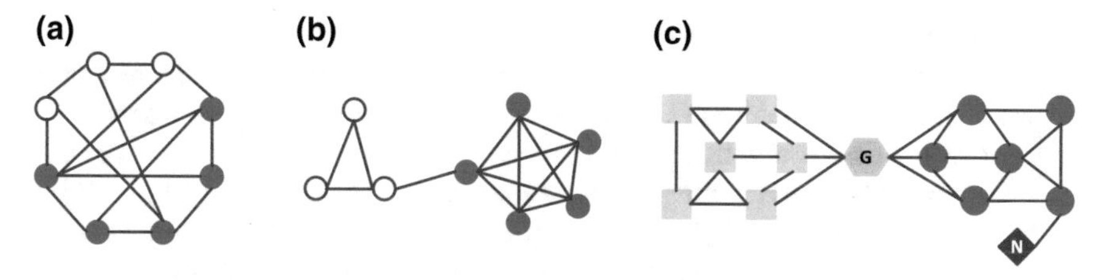

Fig. 3.2 **a** and **b** depict graphs with same density [33]. **c** Output from the SCAN algorithm containing two clusters, a hub G, and an outlier N [72]

$$\sigma(N(v)) = \frac{2|E_v|}{|V|(|V|+1)} \tag{3.14}$$

Given $\sigma(N(v))$ and the highest k-$core$ number associated with v (denoted by k), the weight of vertex v is assigned as $w(v) = k\sigma(N(v))$. This weight boosts neighborhoods with high density and also rewards clusters that exhibit strong "clique-like" structure.

Phase 2: Molecular complex prediction. Equipped with the vertex weights, MCODE finds complexes in a greedy *seed and extend* manner. It starts with the highest weighted vertices as seeds. Following that, the seed is expanded by including neighbors having weight exceeding a user-specified threshold. This parameter is known as the *vertex weight percentage* (VMP) parameter. The expansion stops once no vertices satisfy the threshold parameter and the complex can no longer be grown. The algorithm then proceeds with the next remaining highest weight vertex as new seed.

Phase 3: Post-processing. In this step, complexes are pruned when they do not have at least a 2-$core$. A 'fluff' operation is also introduced to increase the size of complexes. The resulting complexes are then ranked and scored based on their densities.

Clustering techniques that aim to generate clusters solely based on density (e.g., MCODE) may not be able to handle the impact of different graph topologies with same density effectively. For example, consider the two graphs in Fig. 3.2. Observe that the density of both graphs is 0.5. However, intuitively the graph in Fig. 3.2a comprises of a single cluster whereas the one in Fig. 3.2b contains two clusters. Pure density-based clusterings may fail to identify these clusters accurately.

Altaf-Ul-Amin et al. [33] proposed an algorithm called DPClus to address the aforementioned issue by leveraging on the notion of "periphery" to distinguish different graph topologies having same densities. Given a density threshold σ_t and a *cluster property*[1] threshold c_p, DPClus follows a seed and extend strategy to identify dense subgraphs. Similar to MCODE, it first chooses the highest *weighted vertex* as a seed which forms the initial cluster. The *weight* of a vertex u is computed by aggregating the weights of the edges to u where an edge weight is measured by the number of

[1] *Cluster property* of a vertex v with respect to any cluster k of density σ_k and size $|V_k|$ is the ratio of the total number of edges between v and each of the vertices of k to $\sigma_k \times |V_k|$.

common neighbors of the associated proteins. In the case the highest node-weight is zero, the highest degree vertex is considered as a seed. Next, this initial cluster is grown gradually by adding the neighboring vertices that are not yet part of the cluster one at a time sorted by their *priorities*. Given a cluster, a neighbor's *priority* to it is computed by aggregating the vertex weights in the cluster and the number of the edges between the neighbor (not part of the cluster) and the cluster vertices. During the sorting of neighbors, fine tuning may be performed under certain conditions to allow formation of some sparse clusters by checking for periphery using c_p. `DPClus` only adds a vertex in a cluster if the density and cluster property of the cluster are not lower than σ_t and c_p, respectively, after its addition. The final cluster is then removed from the network and the vertex weights of the remainder network are recomputed to generate the next cluster based on the aforementioned steps. This process continues until all edges of the network are exhausted.

The time complexity of `DPClus` is $O(|V|^3)$. Observe that due to recomputation of the weights, the algorithm is expensive and also ignores biological information associated with vertices due to removal of some vertices at each iteration. Additionally, due to removal of a cluster from the original network after each iteration, `DPClus` cannot generate overlapping clusters directly. Hence, it generates overlapping clusters by extending the non-overlapping ones through incorporation of their neighbors in the original network.

Li et al. [35] proposed `IPCA` to address the aforementioned limitations of `DPClus` by exploiting their observation that many protein complexes typically have small diameter and average vertex distance. Hence, in contrast to `DPClus`, it searches for subgraphs having small diameter and whose cluster property is above the *interaction probability* threshold. The *interaction probability* of a vertex v to a subgraph S is defined as the number of edges between v and S normalized by the total number of vertices in S. Observe that it is very similar to c_p in `DPClus`, differing by a factor of σ_k. Similar to `DPClus`, `IPCA` also follows the seed and extend strategy. The vertex weighting step is identical to `DPClus` except that it also sorts the vertices by their weights in a decreasing order and store them in a queue. Similar to the aforementioned seed-expansion techniques, the vertex with highest weight is selected as a seed and is expanded based on the interaction probability of a neighboring vertex without undertaking any fine-tuning to sort the neighbors. Note that a vertex whose interaction probability or diameter is below the user-defined threshold will not be added in the cluster. Once a cluster is generated, all vertices from this cluster is removed from the queue (not from the original network) and the first vertex remaining in the queue is selected for the generation of the subsequent cluster. This process goes on until the queue is empty. Observe that since the vertices are not removed from the original network, in contrast to `DPClus`, the vertex weight computation step is performed only once and generates overlapping clusters directly. This ensures that `IPCA` has a superior running time than `DPClus`.

Another greedy search method is the *Restricted Neighborhood Search Clustering* algorithm (`RNSC`) [36], which deploys a cost-based partitioning algorithm. Specifically, a random clustering is constructed. After that, it moves vertices from one cluster to another iteratively in a randomized manner in order to improve a cost function.

It uses two types of score: the *naive* score and the *scaled* score. The value of the former, for a vertex i, is the sum between the number of neighbours that are not in the same cluster of i, and the number of vertices that are not neighbours of i but belong to the same cluster. The *scaled score* for a vertex i that belong to a cluster C is its naive score normalized by the number of vertices in C and the number of neighbours of i. Note that this algorithm requires as an input the number of clusters to be extracted. Furthermore, even though it was designed specifically for PPI networks, it does not support the detection of overlapping clusters and fails to handle edge weights.

Pei et al. [40] proposed a seed-refine algorithm to detect dense but small subgraphs as clusters by exploiting a novel statistically meaningful subgraph quality measure based on hypergeometric distribution. It iteratively finds a seed subgraph centred on a *seed edge* and refines it until the quality of the *refined subgraph* cannot be further improved. Specifically, an edge that is not part of previously detected refined subgraphs is chosen as a seed edge and then *seeding vertices* (a seeding vertex is connected to both the vertices in a seed edge) with respect to it are located to generate candidate seed subgraphs. Then, the seed subgraph with largest number of vertices is refined by adding or removing vertices until its quality cannot be improved any further. The refinement process not only ensures removal of weakly connected vertices in the subgraph but also controls the overlap between subgraphs (clusters) in the following way. The edges in a refined subgraph are disallowed to appear again in later seed subgraphs but the algorithm does not prevent inclusion of edges in previously refined subgraphs during the refinement process. This ensures the possibility of generating overlapping subgraphs in a controlled manner.

Given the growth of PPI data, it is imperative for clustering techniques to handle large PPI networks in a scalable manner. The `SPICi` method [44] aims to handle the computation complexity of clustering large PPI networks. `SPICi` grows a module, one at a time, from a seed comprising a pair of proteins. To identify a seed, it identifies the node with the highest sum of edge weights connected to the node (*support*) followed by a binned selection process that identifies the best pair of nodes as seed. Following seed selection, a module is formed from the seed by greedily adding an adjacent (unclustered) node with the highest support score. Nodes are added so long the overall module density and/or remaining highest support exceeds their respective *density threshold*. Once a module is identified, the subgraph is removed from the PPI network and the process continues to identify remaining modules. The `SPICi` method has a time complexity $O(VlogV + E)$ and a space complexity $O(E)$, allowing it scale to large PPI networks.

The `ClusterONE` method [19] detects overlapping clusters in a PPI network using a greedy seed and extend heuristic. The approach starts with a single protein node and then greedily adds or removes nodes to find a new group of proteins that shows greatest improvement in cluster cohesiveness (Eq. 3.6). Following that, the extent of overlap among the candidate clusters is evaluated, and cluster merging is performed selectively. This approach demonstrated clustering superiority over a variety of methods, including popular methods such as `RRW`, `RNSC`, `AP`, and `MCL` [19]. An advantage of `ClusterONE` is the ability to not just find overlapping clusters, but also clusters that may be contained in another cluster.

All heuristic-based methods, however, have strong likelihood of converging to a local minimum. On the other hand, these methods generally allow identification of overlapping clusters that better reflect the moonlighting property of proteins (as described in Sect. 3.1.2). These methods also rely purely on topological structure to identify functional clusters. Apart from that, they generate clusterings that are *partial coverage*. A *partial coverage* method finds a set of *locally* dense subgraphs of G, and this set of dense subgraphs does not cover the entire network. Among partial coverage methods, it is generally advantageous to have one which achieves a high coverage score. The advantages of partial coverage methods are similar to those enjoyed by local sequence alignment methods. Clusters obtained are often significantly dense subgraphs, and irrelevant clusters that do not meet the objective function are automatically ignored. Hence, clusterings obtained using partial coverage methods are more amenable to human interpretation.

3.2.2 Flow-Based Algorithms

One of the most widely used graph clustering algorithm is Markov Clustering (MCL) [47]. This approach partitions a PPI network into subgraphs using a flow-based approach. Given a PPI network $G = (V, E)$ with a function $f : E \rightarrow \mathbb{R}$ that gives each pair of proteins their BLAST E-value scores, MCL first constructs a *weight transition matrix* given by:

$$W[i, j] = I((i, j))f(i, j) \tag{3.15}$$

where $I(e)$ is the indicator function such that $I(e) = 1$ if $e \in E$ and $I(e) = 0$ otherwise. Given the weight transition matrix, normalization is performed to obtain the column-wise transition probability matrix:

$$M[i, j] = \frac{W[i, j]}{\sum_x W[i, x]} \tag{3.16}$$

The MCL clustering algorithm simulates the convergence and expansion of flows by iteratively alternating the following two steps until convergence: (1) the *expansion operator* and (2) the *inflation operator*. In the expansion operator, the transition matrix M is raised to the power of m:

$$M_t[i, j] = (M_{t-1}[i, j])^m \tag{3.17}$$

Intuitively, this step can be thought of as transforming M_{t-1} to a transition probability matrix of all random walks over m steps. In the inflation operator, the matrix takes its Hadamard power over $r > 1$ followed by renormalization. This corresponds to an entry-wise power and normalization:

$$\Gamma_r M_t[i, j] = \frac{M_{t-1}[i, j]^r}{\sum_x M_{t-1}[i, x]^r} \tag{3.18}$$

Since entry-wise transition probabilities are raised to a power of $r > 1$, entries with high transition probabilities are favored (i.e., inflated) while entries with low transition probabilities are suppressed, thus favoring densely connected regions.

The MCL approach, however, may generate clusterings with *imbalanced clusters*, where clusters may have significantly different sizes. The occurrence of singleton clusters is one side-effect of having imbalanced clusters. To this end, the MLR-MCL algorithm [48] is proposed to construct more balanced clusters by augmenting the MCL method.

The MCL approach is also a partitioning algorithm that constructs non-overlapping clusters. Another extension of MCL addresses the non-overlapping nature of MCL clusters by introducing a MCL-based clustering strategy that creates overlapping clusters. Here, the authors propose the SR-MCL method [50] that extends the MCL approach by iterative re-execution of the underlying MCL clustering while ensuring the clusters are different. A post-processing is then applied to remove uninformative, redundant clusters, and the final set of overlapping clusters is obtained.

Pereira-Leal et al. [52] preprocess the PPI network into a *line graph* where each vertex represents an edge between a protein pair in the PPI network. These vertices are then connected by edges representing the shared proteins. Figure 3.3 depicts an example of the line graph representation of a PPI network. The advantages of using line graph is that it is more structured than the original network by taking into account the higher–order local neighborhood of interactions. Each edge of the line graph is weighted by averaging the weights of the original interactions. Then, the MCL algorithm is applied on this transformed network to find functional modules. Specifically, the discovered clusters on the line graph are transformed back to the original PPI network.

Since essential proteins have a close connection to overlapping clusters [40], flow-based approaches such as [53] exploit the notion of *informative proteins* for finding overlapping clusters. Specifically, this approach assigns weights to the interactions in a PPI network by quantifying functional relationships between the interacting proteins using *semantic similarity* and *semantic interactivity* measures. These measures are computed using the GO annotations associated with the proteins. Then, the algorithm selects a small number of *informative proteins* based on the weighted degree (sum of the weights of the edges between a vertex and its neighbors) of the proteins, which are

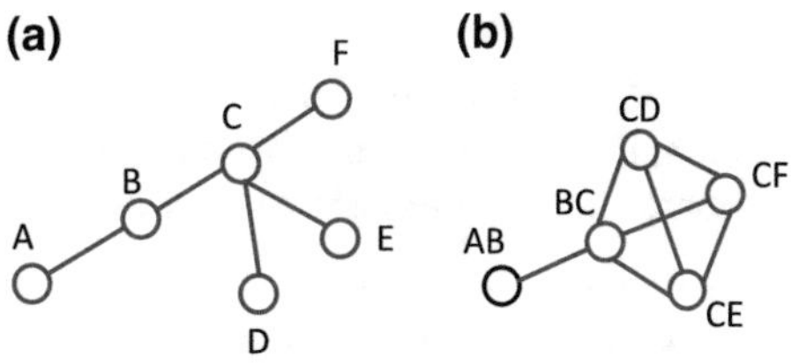

Fig. 3.3 **a** A graph. **b** Line graph of **a** [52]

then used as representatives of the functional modules. Next, from each informative protein s it simulates information flow through the entire network to determine the proteins that are functionally influenced by s. These proteins along with s form a preliminary cluster. Lastly, functionally similar modules are merged to form the final collection of modules. Observe that this approach may generate overlapping modules as it is possible for a set of informative proteins to influence the same protein.

In [54], Cho et al. extended the informative protein-based approach by leveraging the notion of *functional flow pattern*, which is a sequence of functional influence of a source protein to a set of target proteins. Specifically, it discovers a set of functional flow pattern for each module identified in the above steps. Next, a pattern-based clustering algorithm [90] is exploited to identify final modules based on the assumption that two source proteins are likely to have same functions if they have similar functional flow patterns.

A limitation of [53, 54] is that the threshold for merging similar modules iteratively is set manually, which is subjective and needs to be modified according to the underlying data set. `IQ-flow` [55] addresses this issue by integrating the flow-based technique with quantum-behaved PSO (QPSO) [91] so that appropriate merging threshold can be automatically computed. Furthermore, it considers bridging nodes[2] while generating the modules as their exclusion contribute towards more accurate clustering results.

Flow-based clustering methods are full coverage. A limitation of these methods is their inability to utilize the rich information provided by annotations (except for [53, 54]). These annotations can guide the clustering process to identify clusters that are compatible with biological knowledge.

3.2.3 Complete Enumeration Algorithms

Complete enumeration algorithms aim to enumerate all possible subgraphs in G with density exceeding a specified threshold. Several algorithms have been proposed that leverage the notion of clique and its variants.

Spirin and Mirny [56] proposed three techniques for detecting protein complexes and functional modules from PPI networks. The first approach finds cliques as modules by complete enumeration. It begins with $k = 3$ and continue finding cliques with $k > 3$ until no more cliques can be found in the PPI network. Since the network considered by them is sparse, this approach can find clusters relatively quickly. The second approach leverages the notion of *superparamagnetic clustering* (SPC), which assigns to each vertex a "spin" with several states. Since the spins of connected vertices interact, the intuition behind this technique is to detect *correlated* spins as the fluctuation of the spins of vertices in a dense cluster are highly correlated. Lastly, they proposed a Monte Carlo optimization-based technique (MC) where finding highly connected

[2] A bridging node in this work refers to a node having less than 3 degree but is connected to nodes with more than 15 degree.

set of vertices is formulated as an optimization problem. It begins with a randomly selected connected vertices and proceeds by moving selected vertices (according to Metropolis criteria) along the edges of the network to maximize the objective function. The MC-based approach has better performance than SPC for high density networks, whereas SPC excels in finding clusters that have sparse connections to the network. The clusters discovered by the aforementioned techniques may further be cleaned and merged based on their statistical significance.

The `CFinder` method [57] identifies a set of *k-clique modules* in a PPI network where k-cliques correspond to k node complete subgraphs of G with a maximum density of 1. It is based on a deterministic approach called the *Clique Percolation Method* (CPM) [92], which generates overlapping clusters by finding k-clique percolation communities. The algorithm first generates a *clique-clique overlap matrix* by extracting all cliques from the PPI network. This matrix is then used to identify k-clique communities by setting to 0 all diagonal entries in the matrix with value less than k and all off-diagonal entries with value less than $k-1$. Following that, connected components in the matrix are identified as a k-clique community.

Zhang et al. [58] also extended CPM by using a different rule to detect clusters. Given a user-defined parameter S, first they generate initial clusters using $k = 3$. After that they iteratively use $k + 1$ to separate clusters of size larger than S until all modules having size less than S are obtained. However, a limitation of this approach is the restrictive assumption that all modules have 3-clique topological property. This may not be necessarily true for some modules (e.g., spoke-like modules).

More recently, Cui et al. [59] showed on the yeast PPI network that *near-cliques* may reveal better quality functional modules compared to overlapping cliques. Specifically, they consider three types of near-cliques as depicted in Fig. 3.4 and proposed an efficient heuristic algorithm to identify them. First, it searches for the A and C categories of near-cliques in G satisfying any one of the following properties: (a) for each vertex x in a near-clique g', the number of edges connecting x to other vertices (i.e., $indegree(x)$) in g' must be greater than or equal to the corresponding number of edges connecting x to other vertices that are not in g'; (b) the $indegree(x)$ of a vertex x in g' must at least be equal to half the number of vertices in g'. After that, it assigns the *clique index* to every vertex in it and if necessary merges two cliques into a near-clique of *Type B* category. If a vertex x is outside a clique and forms the structure of *Type A* category, then the aforementioned conditions are checked for the formation of *Type A* near-clique and x is assigned the corresponding clique index. Similarly, near-cliques of *Type C* categories are formed when there is a common

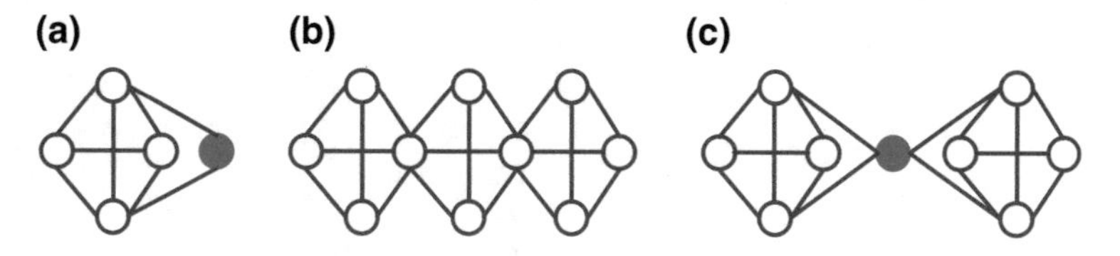

Fig. 3.4 Types of near cliques. **a** Type A. **b** Type B. **c** Type C [59]

protein, which is not a member of any clique, having two or more interactions with two or more cliques. In this algorithm, the size of the near-cliques are controlled by a user-defined parameter.

Extending the idea of clique enumeration to more general graphs, the DME method [64] enumerates *all* subgraphs that satisfy a minimum density threshold (i.e., modules). This approach models the search process using a tree. The root of the tree is an empty set, while any children node in the tree is a superset of that node's parent, and for any path from the root to the leaf, the *module density* is monotonically decreasing. Here, the *module density* refers to the average pairwise weights of the edges in a module. By enforcing density guarantee in the search tree, the DME method prunes all unnecessary explorations during the search process while exhaustively enumerating all sufficiently dense modules. Although the DME method can find all dense modules, it is computationally expensive. As such, it is applicable to only relatively small networks.

The Clustering-based on Maximal Cliques (CMC) [61] method assesses the probability that two proteins are in the same protein complex using an iterative scoring algorithm followed by a maximal clique finding process. It first generates all maximal cliques of a PPI network; this is followed by a series of steps that merges highly overlapping cliques. This approach yields a set of densely interacting cliques that are fairly non-redundant. It is also shown to be less sensitive to parameters compared to flow-based techniques such as MCL [47]. An weakness of CMC, however, is that it identifies only clusters that correspond to a clique topology and cannot handle edge weights.

A common theme among complete enumeration algorithms is exhaustive search. While such search enables identification of all relevant modules within a PPI network, it is computationally expensive. Therefore, their applications are limited to relatively small PPI networks.

3.2.4 Random Walks and Message Passing Algorithms

A well-known strategy to cluster a PPI network is to model the graph as a random walk model, and then after performing a series of random walks, identify a set of clusters.

Nibble [93] is an approach that relies on a modified random walk strategy [17] and finds clusters with low conductance (Eq. 3.7). A strong advantage of this method is its scalability—Nibble runs in nearly-linear time in the size of the cluster outputs. Also, Nibble defines the following vectors defined on a vertex set S:

$$\chi_S(u) = \begin{cases} 1 & \text{for } u \in S \\ 0 & \text{otherwise} \end{cases} \tag{3.19}$$

$$\varphi_S(u) = \begin{cases} k_u/mu(S) & \text{for } u \in S \\ 0 & \text{otherwise} \end{cases} \tag{3.20}$$

Given a graph G, `Nibble` first constructs its adjacency matrix A:

$$A(u, v) = \begin{cases} 1 & \text{if } (u, v) \in E, u \neq v \\ k & \text{if } u = v \text{ and u has k self loops} \\ 0 & \text{otherwise} \end{cases} \quad (3.21)$$

The random walk matrix is defined as $M = (AD^{-1} + I)/2$ where D is the diagonal matrix of node degrees. The distribution of the random walk given the start seed v after t steps is then $p_t = M^t \chi_v$. `Nibble` also introduces the *truncation* operation on p_t given by:

$$[p]_\varepsilon(u) = \begin{cases} p(u) & \text{if } p(u) \geq k_u \varepsilon \\ 0 & \text{otherwise} \end{cases} \quad (3.22)$$

which truncates every $q_t(u)$ less than $d(u)\varepsilon$ to zero.

`Nibble` runs in iterations. Starting at a seed vertex v, at each iteration, a random walk is performed followed by a truncation operation. After a few steps, the distribution of the truncated random walk $[p_t]_\varepsilon$ can be used to identify a cut with low conductance. If a clustering with desirable clustering score occurs in one of the steps, the algorithm terminates and the clustering is set as output. Otherwise, the iterative procedure is continued until a predefined maximum number of steps is reached. In this case, no desirable clustering is obtained.

The Repeated Random Walk (`RRW`) method [68] clusters PPI networks using a random walk with restart methodology. The basic idea behind `RRW` is the following. Given a cluster of nodes, the algorithm tries to expand it with the aim of including proteins with high proximity to that cluster. An advantage of the `RRW` approach is its ability to admit overlapping clusters. First, `RRW` constructs the transition matrix P from $G = (V, E)$ and edge weight function $f : E \rightarrow \mathbb{R}$. Then, for each $v \in V$, it computes the stationary distribution vector associated with v as starting node, defined by:

$$m[v] = \alpha s + (1 - \alpha) P^T m[v] \quad (3.23)$$

where s is the start vector such that v is the starting node; and α is the *restart probability* parameter, which defines the probability that the walk restarts at the starting vector s. Additionally, the stationary distribution x_C of a set of proteins $C = \{v_1, v_2, \ldots, c_k\}$ is given by $x_C = \sum_{v \in C} m[v]$.

The `RRW` algorithm then proceeds as follows: (1) For each $v \in V$, set $C = \{v\}$, and expand C by identifying proteins that exhibit strong transition probabilities from x_C and adding them to C. The expansion terminates if the next added protein score is below λ percentage of the previously added protein. (2) Given C for each $v \in V$, post-processing is performed to remove highly overlapping clusters.

An approach that is similar to random walk is the message-passing based *Affinity Propagation* (`AP`) method [67]. Affinity Propagation aims to learn a set of centers from the input network such that the sum of squared errors between each input object and its nearest center is minimum. Specifically, these centers are referred to as *exemplars*, which are representative proteins of clusters. First, every protein $v \in V$

is flagged as exemplars. For every protein $i \in V$ and an exemplar $v \in V$, let $r(i, v)$ be the *responsibility* of v given i. Intuitively, responsibility reflects how likely v is the exemplar of i. It is defined as follows:

$$r(i, v) = s(i, v) - max_{u \neq v}\{a(i, u) + s(i, u)\} \tag{3.24}$$

In the above equation, $a(i, u)$ refers to *availability* and is defined as follows:

$$a(i, v) = min\{r(v, v) + \sum_{u \in \{u | u \notin \{i, v\}\}} r(u, v)\} \tag{3.25}$$

Initially, all availabilities are set to zero. Messages in form of availabilities and responsibilities are then passed among neighbors and exemplars until convergence is achieved. It has been shown that `AP` underperformed `MCL` in clustering of PPI networks [94].

While both `AP` and `Nibble` do not admit overlapping clusters, the `RRW` method is one random walk-based method that allows overlapping clusters. Note that the aforementioned techniques are purely topology-driven clustering methods. None of these methods utilize the wealth of annotation data to compute important functional clusters. Typically, random walk and message-passing based clustering methods construct a partitioning on the PPI network, implying that the clustering is full coverage.

3.2.5 Graph-Cut and Hierarchical Clustering Algorithms

Hierarchical clustering is a popular technique for clustering PPI networks due to modular nature of such networks [8]. Intuitively, hierarchical clustering algorithms view the inherent hierarchy of a network in the form of a tree. Hence, these algorithms can be classified into the following two types based on the way the hierarchical tree is constructed. The *agglomerative* algorithms take a top-down approach by first constructing each vertex as a cluster (root) and then merge these vertices iteratively. Specifically, it calculates a weight to measure how closely connected two groups of vertices are and links are added iteratively between them in order of decreasing weight. On the other hand, the *divisive* algorithms take a bottom-up approach by recursively splitting the network into two or more subgraphs These algorithms employ various heuristic rules (e.g., edge betweenness, minimum cut) to merge vertices or divide networks. In this section, we discuss these representative techniques.

3.2.5.1 Betweenness-Based Clustering

Chen et al. [69] extended `G-N` algorithm [95] for clustering in weighted PPI network. In this approach, edges of the network are weighted using microarray datasets where an edge weight represents the dissimilarity between expression profiles of a pair of genes. The shortest paths are then computed based on these edge weights. Furthermore, as the G-N algorithm may generate unbalanced partitions under certain scenario, the authors exploited a modified definition of edge betweenness (Eq. 3.11) to compute the clusters. Specifically, shortest paths with distinct end points are only considered for computing edge betweenness and is implemented using the Maximum Bipartite Matching algorithm.

Luo et al. [5] developed an agglomerative algorithm called MoNet that redefines the concept of module by extending the notion of degrees from individual vertices as proposed by [4] (Eqs. 3.1 and 3.2) to subgraphs. A subgraph of a PPI network is considered a module if its *modularity* is greater than one. In contrast to the classical notion of modularity (Eq. 3.13), the authors define the *modularity* of a subgraph g' as the ratio of its *indegree* to *outdegree* where the *indegree* of g' is the number of edges within it and its *outdegree* is the number of edges which are connected to vertices outside of g'. Based on this module definition, MoNet finds *simple* modules from a PPI network. It first initializes each vertex in the network as a mergable singleton subgraph with no edges. Then these clusters are *merged* iteratively into modules by adding edges to the clusters in the reverse order of deletion as followed by the G-N algorithm [95]. Note that the G-N algorithm orders the edges in descending order of their edge betweenness values. Furthermore, lower the betweenness value of an edge the more likely it is part of a module [95]. The merging is performed either between two non-modules or between a module and a non-module. The time complexity of this algorithm is $O(|E|^2|V|)$.

Due to the hierarchical nature of the betweenness-based clustering algorithms, these techniques cannot be used to detect overlapping clusters. Furthermore, these clustering approaches are computationally expensive as each edge needs to be repeatedly evaluated.

3.2.5.2 Shared Neighbor-Based Clustering

Betweenness-based clustering techniques are expensive especially for large PPI networks as they typically have quadratic running time. Mete et al. [72] aim to alleviate this problem by proposing a linear time algorithm called SCAN that exploits the notion of *indirect connections* for detecting modules as well as hubs and outliers in PPI networks. The basic idea behind this algorithm is that the similarity between two vertices can be measured by the number of neighbors they share. Then, vertices whose similarity is beyond certain threshold can be assigned to the same cluster. Observe that this notion of indirect connections to determine cluster membership departs from the traditional clustering strategies based on the number of edges within or outside clusters (Eqs. 3.1 and 3.2). Given a similarity threshold ε and a cluster size threshold μ, SCAN performs a single pass on G and classifies each vertex either a member of a cluster or otherwise. For each unclassified vertex v_i, it checks if it is a *core*. A *core* vertex has at least μ neighbors with *structural similarity*[3] greater than or equal to ε. In the case v_i is a core, it is expanded to form a new cluster by adding ε-neighbors of v_i. If a vertex is not a core, then it is classified as a non-member. Lastly, among non-member vertices, if a vertex has edges to two or more clusters, it is marked as a hub. Otherwise, it is an outlier. Figure 3.2c depicts an example of clusters, hub, and outlier generated by `SCAN` on a toy network.

Similarly, the `NeMo` algorithm [20] predicts the association between a pair of proteins based on the idea of shared neighbors—two proteins are deemed highly connected if they share most of their neighbors. Specifically, it uses the log odds scores of protein pairs (Eq. 3.9) to this end. Given these scores, `NeMo` then proceeds to identify clusters using a hierarchical agglomerative clustering approach. Both *complete-linkage* and *single-linkage* strategy are considered. Node pairs are processed greedily based on their log odds scores. `NeMo` only groups pairs having expected number of shared neighbors greater than by chance. The greater the fraction of shared neighbors between two proteins a and b, the larger their log odd score. A *collapse* procedure is also introduced to prune insignificant structures from the result. Given the hierarchical tree T, any internal node p having children m and n such that n is a leaf node and m is an internal node, `NeMo` collapses the edge between p and m.

3.2.5.3 Annotation-Driven Clustering

Typically, a clustering is obtained from a hierarchical clustering tree by “cutting” the tree at a particular level. For instance, given a binary tree of five levels with 2^5 leaf nodes, a clustering with 4 clusters can be obtained by grouping the leaf nodes by their level 3 ancestors. The core idea of `Tree-Sniping` [71] is the following proposition: rather than cutting at a single level, snip the tree at selected edges at different levels. With the added flexibility, `Tree-Sniping` can pick and choose snips that maximizes the compatibility of the clusters with its constituent proteins' annotation.

Let T be the hierarchical clustering tree obtained from a graph clustering of G. For each node $v \in T$ and *snips* l and k, let $minMis(v, l, k)$ be the minimal number of misclassified leaves (protein label not compatible with cluster label) when v is labeled as l and there are k snips in the subtree rooted at v. Also, let $minNum(v, k) = \min_l minMis(v, l, k)$. Then for each k and l, $minMis$ is defined as the following recursive function:

[3] The structural similarity of a pair of vertices is measured by normalizing the number of common neighbors with the geometric mean of the two neighborhoods' size.

$$minMis(v,l,k) = min \begin{cases} minMis(left,l,r) + minMis(right,l,k-r) \\ \quad 0 \le r \le k \\ minMis(left,l,r) + minNum(right,k-r-1) \\ \quad 0 \le r \le k-1 \\ minNum(left,r) + minMis(right,l,k-r-1) \\ \quad 0 \le r \le k-1 \\ minNum(left,r) + minNum(right,k-r-2) \\ \quad 0 \le r \le k-2 \end{cases} \tag{3.26}$$

To this end, the recursive function is solved via dynamic programming method that traverses the tree in a bottom-up manner. The `Tree-Sniping` method does not scale well with the dimensionality of the labels. Given the high dimensionality of GO annotations, the misclassified labels will dominate the scores and mask the relatively fewer conserved labels. Experiments described in [71] are applied on one to three labeled genes. For instance, only three GO terms are manually selected for the clustering experiments. `Tree-Sniping` also performs best when the annotations largely form a partition. Consider for example the biological process GO term and a PPI network labeled with biological process-related GO annotations. With `Tree-Sniping`, no proteins are considered misclassified under this overarching GO term, and as such the single large cluster associated with a biological process is considered the optimal solution.

Similar to `Tree-Sniping`, `VI-Cut` [23] is a tree-sniping approach that relies on the *variational information* metric to generate clusters that "match" with MIPS annotations of the proteins. Suppose one wishes to cluster a graph $G = (V, E)$ with annotations D. The `VI-Cut` algorithm first obtains a hierarchical clustering tree T on G as input (this tree can be obtained using any hierarchical clustering techniques). Given T, `VI-Cut` determines the cuts of the tree such that the variational information distance measure $VI(C, D)$ (Eq. 3.12) is minimized so that the generated clusters agree well with the proteins with GO annotations in D. The authors show that $VI(C, D)$ is equivalent to:

$$VI(C,D) = \sum_{x \in C} q(x) \tag{3.27}$$

where $q(x) = p(x)\log p(x) - 2\sum_{d \in D} p(x,d)\log p(x,d)$. Here, x denotes a node in the hierarchical decomposition tree and $p(x)$ is the probability that a protein with an annotation belongs to x. Also, $p(x, d)$ is the joint probability that protein with an annotation belongs to x and has annotation d. Any cut that is made should minimize $VI(C, D)$. `VI-Cut` computes this via dynamic programming by solving the following recursive problem:

$$CutDist(x) = min \begin{cases} q(x) \\ \sum_{y \in \text{Children}(x)} CutDist(y) \end{cases} \tag{3.28}$$

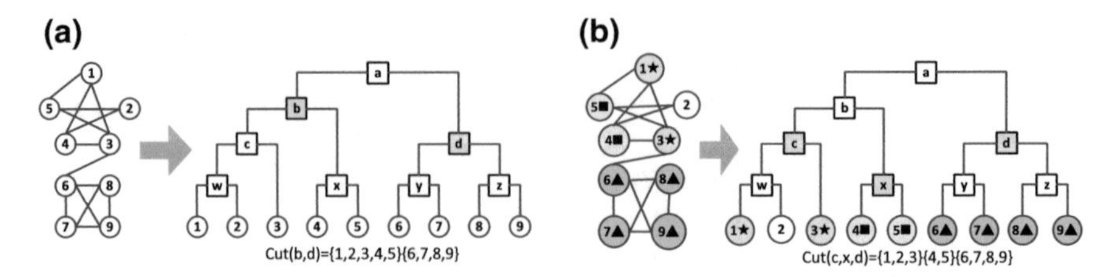

Fig. 3.5 The VI-Cut algorithm illustrated with a toy example [23]

Figure 3.5 demonstrates the `VI-Cut` algorithm. Consider the network in Fig. 3.5a. learly, it consists of two dense subgraphs and most topology-based hierarchical clustering approaches will identify them and generate a decomposition as depicted in the right hand side of Fig. 3.5a. That is, proteins {1, 2, 3, 4, 5} and {6, 7, 8, 9} form two separate clusters. However, if annotations associated with some of the proteins are known (shown by symbols), then the cut {b, d} is not a good solution as it groups proteins {1, 3} and {4, 5} having disparate annotations together. Starting from the root of the hierarchical clustering tree (cluster of all proteins), the `VI-Cut` algorithm incorporates this intuition by generating cuts that ensure the match between the clusterings and the annotations are as close as possible. Consequently, it chooses the cut {c, x, d} as the best solution, which induces clusters {1, 2, 3}, {4, 5}, and {6, 7, 8, 9} (Fig. 3.5b). However, this approach scales poorly with the dimensionality of the annotations per protein, which makes it unsuitable for richly attributed GO annotation data.

A strength of hierarchical clustering methods is the potential of constructing a hierarchy of clusterings imitating the hierarchical organization of PPI networks. This allows analysis of clusterings at multiple levels of granularity. These methods are also full coverage. However, as mentioned earlier such clustering does not admit overlapping clusters naturally without further preprocessing. Furthermore, it is also known that they are sensitive to the noisiness of PPI networks [53].

3.2.6 Multiple Clustering-Based Algorithms

Asur et al. [21] proposed ensemble clustering framework where a range of independent clusterings are obtained and *combined* to construct a single *consensus clustering*. The intuition of ensemble clustering is that the combined clustering may yield high confidence clustering even in the presence of noise. In this approach, six base clusterings are first constructed based on three traditional graph partitioning algorithms (repeated bisections, direct k-way partitioning, and multilevel k-way partitioning) and two topology-driven distance metrics (based on clustering coefficient and edge betweenness (Eqs. 3.3 and 3.11)). Then a PCA-based consensus method is deployed for dimension reduction of the ensemble clustering problem. They also adapted the design to support soft ensemble clustering.

Greene et al. [74] proposed a non-negative matrix factorization-based (NMF) ensemble framework for clustering PPI networks. Observe that since NMF can be used to identify overlapping structures, this approach is particularly suitable for soft clustering. In the *generation* phase, a collection of NMF factorizations is produced as base clusterings. In the *integration* phase, these factorizations are combined using the min-max hierarchical clustering technique to form a meta-clustering. Lastly, a soft hierarchy is generated from the meta-clustering and redundant tree nodes are recursively eliminated (if necessary).

The `MOD-ILP` algorithm [24] casts the PPI network clustering as an Integer Linear Program (ILP) problem. In `MOD-ILP`, the *modularity* of a subgraph of G (Eq. 3.13) is proposed to measure the clustering objective score of a clustering. The ILP problem is then posed as finding the clustering S that maximizes the modularity objective function:

$$max \sum_{u,v \in V} \frac{I_{uv} - k_u k_v / 2|E|}{1 - x_{uv}} \tag{3.29}$$

$$s.t. \quad x_{uv} + x_{vw} \leq x_{uw} \qquad \forall u, v, w \in V \tag{3.30}$$

$$x_{uv} \in \{0, 1\} \tag{3.31}$$

The first constraint enforces the transitive property of cluster membership, while the second constraint arises from the combinatorial nature of the problem. An interesting novelty of `MOD-ILP` is its ability to general an ensemble of clusterings, as opposed to other clustering methods that produce only a single clustering of G. The above formulation, however, only admits one possible clustering of the network. Suppose the first clustering obtained is S_0. `MOD-ILP` iteratively generates a new clustering S_t where $t > 0$ from the past clustering S_{t-1} by imposing a "uniqueness" constraint, which forces the next set of results to be different:

$$S_{t-1} \cdot (1 - S_t) \leq d^0_{merge} \tag{3.32}$$

$$(1 - S_{t-1}) \cdot S_t \leq d^0_{split} \tag{3.33}$$

where d^0_{merge} and d^0_{split} are real-valued parameters that define the degree of difference required. Using the above formulation, `MOD-ILP` constructs a set of clusterings that can be seen as an ensemble of near-optimal solutions, and these ensembles were utilized to study the robustness of the modularity landscape.

3.2.7 Genomic Data-Driven Clustering Algorithms

The aforementioned techniques of PPI clustering mostly leverage graph theoretic properties to find clusters and only depend on the PPI network. Unfortunately, recall that PPI networks are noisy with both false positive and false negative interactions. Hence, in order to generate robust clustering techniques it is imperative to improve the reliability of the interaction data. A popular strategy adopted in the literature to tackle this issue is by integrating genomic and proteomic data, GO annotations, etc. with PPI. We have already discussed how GO annotations are leveraged in annotation-driven hierarchical clustering techniques. In this section, we briefly introduce representative efforts in integrating genomic or proteomic data to create robust clustering.

Segal et al. [75] introduced a unified probabilistic model to detect functional modules from gene expression and PPI data based on the assumptions that (a) genes in the same pathway display similar expression profiles and (b) protein products of genes that work together to accomplish certain task often interact. They first cluster the expression data to find pathways and then build a probabilistic model by integrating Naive Bayes model of the gene expression component and Markov random fields model of the interaction data. This unified model is trained using the EM algorithm to learn the parameters of the model.

Since interacting proteins tend to be localized in the same subcellular component and demonstrate similar expression profiles, Lu et al. [78] integrated PPI data with subcellular localization and expression data to identify functional modules. Specifically, they devised a hierarchical clustering strategy that computes proximity between two protein groups by leveraging spatial and temporal information from localization and expression data, respectively, in addition to the interaction information. This ensures that co-localized and co-expressed protein groups are clustered first during hierarchical clustering tree construction.

Maraziotis et al. [81] proposed an algorithm called DMSP that also finds functional modules by integrating gene expression and PPI data. Specifically, the gene expression profiles are first clustered using the fuzzy c-means algorithm, which are used to assign weights to the edges of the PPI network. Particularly, the centroids of the clusters of the corresponding pair of genes associated with an interacting protein pair are used to compute the weight of an interaction by aggregating the distance of each gene from its centroid and the distance between the centroids. Then the algorithm unveils the functional modules in two phases by expanding the *kernel neighborhood* from a seed protein. In the first phase, a subset of the neighbors of a seed protein, which is considered most "promising", is selected as kernel neighborhood based on the density of the kernel as well as its weighted internal and external degrees. In the second phase, adjacent vertices are added iteratively to the selected kernel based on certain criteria related to the number of neighbors of the specific protein and weights of the edges.

Zheng et al. [83] combines seven genomic and four experimental interaction data sets to construct a Bayesian network, which can be used for integrating information from disparate sources. It is then used to compute for each protein pair a likelihood-ratio-based score. Given a threshold for this score, a set of high-confidence protein pairs can be inferred to construct the PPI network. The protein complexes from this inferred PPI network are detected by running MCL on it. However, as they employ MCL, it suffers from the same limitations as discussed earlier.

Ulitsky and Shamir [87] reformulated the problem of finding modules with high confidence connectivity as finding subgraphs satisfying a weight threshold of their minimum cut. To this end, they proposed a probabilistic model and an algorithm called CEZANNE to identify functionally *coherent co-expressed gene sets* by exploiting expression profiles and confidence-scored protein interactions. The goal is to identify *q-connected* modules having maximum co-expression score. Informally, a set of vertices $V' \subseteq V$ is *q-connected* if for all subset of vertices in V', the probability of at least one edge connecting it with a vertex set not in V' is at least q. Given a PPI network, it first identifies a non-overlapping collection of initial seeds using the MATISSE [96] methodology. Then, it assigns a confidence weight of $-log(1 - p(e))$ to each edge e where $p(e)$ is the probability that e exists. Next, it identifies all disjoint q-connected seeds by finding the minimum cut having weight greater than a threshold T recursively. This cut is used to split the initial seeds. Lastly, the set of initial seeds are optimized (while maintaining the q-connectedness and the threshold T) by adding/removing vertices from a module, reassigning a vertex from one module to another, or by merging two modules. Statistically significant modules are then identified by filtering them based on their p-values.

More recently, Shi et al. [89] proposed a neural network-based semi-supervised learning method that leverages proteomic features (protein length, polarity of amino acids) of subgraphs in a weighted PPI network along with their topological features (e.g., clustering coefficient, edge weight statistics, degree statistics) to generate protein complexes. Given a training set of protein complexes (from MIPS protein complexes) and randomly generated non-protein complexes, it first uses a two-layers feed-forward neural network model to train the parameters for the prediction model by utilizing the topological and proteomic features of these complexes. The model is then utilized to select protein complexes by using a seed-expansion strategy where an initial set of seed nodes are expanded by adding neighboring proteins until no more proteins can be added. These newly generated complexes are fed back to the prediction model to recursively update the parameters and find new complexes.

3.3 Cluster Validation Measures

Biological validation of the clusters predicted by different clustering techniques is paramount. This is more so because different algorithms may generate different clusters from the same PPI network. Furthermore, a specific algorithm may generate different clusters based on different parameter settings. Consequently, it is important to biologically validate the outcome of a specific algorithm to determine its appropriateness. In this section, we review representative measures for validating the output of PPI clustering techniques.

3.3.1 Functional Homogeneity-Based Validation

Since proteins in a cluster often have similar function, functional homogeneity of proteins in a cluster can be compared with known function annotation (obtained from MIPS or GO) to measure the goodness of the predicted cluster. In the literature, p-value, *clustering score*, and *functional coherence* are some of the key measures used to quantify it. In this section, we briefly discuss these measures in turn.

P-value enables us to calculate the statistical and biological significance of a group of proteins. Informally, it represents the chance of seeing the group of proteins and is computed using hypergeometric distribution. Formally, it is defined as follows:

$$P(C) = 1 - \sum_{i=0}^{k-1} \frac{\binom{|W|}{i}\binom{|V|-|W|}{|C|-i}}{\binom{|V|}{|C|}} \tag{3.34}$$

In the above equation, the predicted cluster C consists of k proteins belonging to the functional group W. Note that smaller the p-value, the more likely the predicted cluster is not randomly formed and hence is biologically more significant.

Although the p-value in the above equation can quantify the quality of a single cluster, it is insufficient to quantify the overall quality of all predicted clusters. Hence, *clustering score* [21, 72] is often used to quantify the quality of all predicted clusters. Let n_S and n_I denote the number of *significant* and *insignificant* clusters, respectively,[4] and $min(p_i)$ be the smallest p-value of the significant clusters. Then, *clustering score* is defined as follows.

$$Score = 1 - \frac{\sum_{i=1}^{n_S} min(p_i) + (n_I \times cutoff)}{(n_S + n_I) \times cutoff} \tag{3.35}$$

Redundancy [52] is another measure for functional homogeneity and defined as follows. Let n be the number of classes in the classification scheme and p_s be the relative frequency of the class in the predicted cluster. Then,

$$R = 1 - \left[\frac{-\left(\sum_{s=1}^{n} p_s \log_2 P_s\right)}{\log_2 n}\right] \tag{3.36}$$

Intuitively, clusters that have many proteins with highly consistent classifications will tend to receive high redundancy scores. On the other hand, if clusters contain many proteins with diverse classifications then they will have low R value.

In [93], the notion of *functional coherence* is used to quantify the functional relatedness of proteins within a cluster and to other proteins in the PPI network. To this end, two types of measures are proposed for a predicted cluster, namely, *absolute* and *relative functional coherence*. The former measures the difference between the average *functional distance* of a protein pair in the network and the average pairwise

[4] A cluster is considered *significant* if its $min(p_i) < cutoff$. Otherwise, it is an *insignificant* cluster.

functional distance of proteins in the predicted cluster C. Intuitively, the functional distance of proteins u and v (denoted as $f(u, v)$) is the number of gene pairs with common LCAs of these proteins in the GO hierarchy. Hence, the absolute functional coherence can be quantified by the following equation:

$$abs\_coh(C) = \frac{\sum_{u \neq v \in V} d(u, v)}{|V|(|V| - 1)} - \frac{\sum_{u \neq v \in C} d(u, v)}{|C|(|C| - 1)} \tag{3.37}$$

where $d(u, v) = \log_{10}(f(u, v))$. The *relative functional coherence*, on the other hand, measures the difference in average functional distance of inter-community and intra-community protein pairs:

$$rel\_coh(C) = \frac{\sum_{u \in C, v \in V - C} d(u, v)}{|C|(|V - C|)} - \frac{\sum_{u \neq v \in C} d(u, v)}{|C|(|C| - 1)} \tag{3.38}$$

3.3.2 MIPS-based Validation

A popular strategy to quantify the quality of a predicted cluster is to compare it with known complexes catalogued in the MIPS database. This is computed using the *overlap score* $O(C_p, C_k)$ [25, 33] between a predicted cluster C_p and a known complex C_k:

$$O(C_p, C_k) = \frac{|V_p \cap V_k|^2}{|V_p| \times |V_k|} \tag{3.39}$$

C_k and C_p are considered a match if $O(C_p, C_k) > \delta$. Based on this overlap score, *sensitivity* and *specificity* measures [25] can be used to compute how the known and predicted clusters are matched. In particular, *sensitivity* measures the fraction of true-positive predictions out of all true predictions. On the other hand, *specificity* is the fraction of true-positive predictions out of all positive predictions. Formally, let TP be the number of the predicted clusters matched by the known complexes with $O(C_p, C_k) \geq \delta$, FP denotes the total number of the predicted clusters excluding TP, and FN be the number of the known complexes that are not matched by the predicted clusters. Then, sensitivity and specificity are computed as $S_n = TP/(TP + FN)$ and $S_p = TP/(TP + FP)$, respectively. Note that these two measures can be combined using F-measure. Apart from these, measures such as positive predictive value (PPV), contingency table, accuracy, and separation are also proposed in the literature to quantify the match between known complexes and predicted clusters [97].

3.3.3 *Other Measures*

In addition to the above measures, the cluster quality can be measured using other dimensions such as the Czekanowski-Dice distance (Eq. 3.10), degree of overlapping clusters, coverage of the clusters, and modularity (Eq. 3.13). For instance, the *overlap* metric can be used to measure the average number of clusters a protein belongs to and hence can be used to quantify the degree of overlap among the clusters. The *coverage* metric can be used to evaluate the fraction of vertices in a PPI network covered by the predicted clusters. It can be computed as the ratio of the total number of proteins in the clusters over the total number of proteins in the PPI network. Clearly, a clustering technique that generates clusters with high coverage is more desirable than the one with low coverage as the former is more representative of the underlying network. Lastly, topology-based metrics such as modularity can be used to evaluate if the predicted clusters are clique-like. However, this measure cannot evaluate the functional coherence of the proteins in a cluster as it ignores annotations associated with vertices.

3.4 Comparative Summary

Table 3.2 summarizes the aforementioned PPI network clustering approaches based on the following properties:

- *Weighted*: It indicates whether the algorithm considers the weights of the edges during clustering.
- *Overlapping*: It indicates whether the algorithm can identify overlapping modules or otherwise.
- *Scalability*: It qualitatively suggests the algorithm's capability to scale to larger PPI networks.
- *Exhaustive*: It indicates whether the algorithm can identify all modules in a PPI network that satisfy the clustering criteria specified by the algorithm.
- *Complete Coverage*: It identifies algorithms that can form clusters on all nodes in a PPI network (i.e., full coverage).
- *Annotation*: It specifies whether an algorithm is annotation aware, that is, whether the clustering process is guided by annotation information encoded within PPI networks.

It is generally desirable for a PPI network clustering algorithm to be exhaustive, scalable, and annotation-aware. It should also admit overlapping clusters (while controlling redundancies), consider edge weights, and admit clustering with complete coverage. Unfortunately, observe that no single algorithm in the literature enjoys all of the above strengths. However, it is worth noting that the performances of these algorithms cannot be summarized by only considering these properties alone. Many algorithms have their own unique characteristics and strengths as remarked earlier.

Table 3.2 Summary of PPI network clustering techniques

Algorithms		Weighted	Overlapping	Scalability	Exhaustive	Full coverage	Annotation
MCODE	[25]	No	Yes	Medium	No	Low	No
DPClus	[33]	No	Yes	Low	No	High	No
IPCA	[35]	No	Yes	Medium	No	High	No
RNSC	[36]	No	No	Medium	No	High	No
Pei et al.	[40]	No	Yes	Medium	No	Low	No
ClusterONE	[19]	No	Yes	Medium	No	No	No
SPICi	[44]	Yes	No	High	No	Yes	No
SPC & MC	[56]	Yes	No	Medium	Yes	No	No
CFinder	[57]	No	Yes	Low	Yes	No	No
Zhang et al.	[58]	No	Yes	Low	Yes	No	No
Cui et al.	[59]	No	Yes	Low	Yes	No	No
CMC	[61]	No	Yes	Low	No	Yes	No
DME	[64]	No	Yes	Low	Yes	No	No
AP	[67]	No	No	Medium	No	Yes	No
Nibble	[17]	No	No	High	No	Yes	No
RRW	[68]	No	Yes	Medium	No	Yes	No
TRIBE-MCL	[47]	Yes	No	Medium	No	Yes	No
MLR-MCL	[48]	Yes	No	Medium	No	Yes	No
SR-MCL	[50]	Yes	Yes	Medium	No	Yes	No
Pereira-Leal et al.	[52]	Yes	No	Medium	No	Yes	No
Cho et al.	[53]	Yes	Yes	Low	No	Yes	Partially
Cho et al.	[54]	Yes	Yes	Low	No	Yes	Partially
IQ-flow	[55]	Yes	Yes	Medium	No	Yes	Partially
MoNet	[5]	No	No	Low	No	Yes	No
Chen et al.	[69]	Yes	No	Low	No	Yes	No
Tree-Snipping	[71]	No	No	Medium	No	Yes	Yes
VI-Cut	[23]	No	No	Medium	No	Yes	Yes
SCAN	[72]	No	Yes	Medium	No	No	No
NeMo	[20]	No	No	Medium	No	No	No
Asur et al.	[21]	No	No	Low	No	Yes	No
Greene et al.	[74]	No	Yes	Low	No	Yes	No
MOD-ILP	[24]	No	No	Medium	No	Yes	No
Segal et al.	[75]	Yes	No	Low	No	No	No
Lu et al.	[78]	Yes	No	Medium	No	Yes	No
Maraziotis et al.	[81]	Yes	No	Low	No	No	No
Zheng et al.	[83]	Yes	No	Low	No	Yes	No
CEZANNE	[87]	Yes	No	Low	No	Yes	No
Shi et al.	[89]	Yes	Yes	Low	No	Yes	No

For instance, the NeMo algorithm is notable for its ability to incorporate indirect interaction information based on shared neighbors. ClusterONE is notable for outperforming many well established methods in reconstructing known complexes. CEZANNE was designed to handle the noisiness of PPI networks by integrating additional data sources.

Additionally, another important limitation of most existing PPI network clustering methods is the emphasis on cluster density in their clustering objective function. For instance, MOD-ILP defines an objective function to maximize the structural modularity of clusters. The MCL-based approaches utilize a sequence of expansion and inflation steps that results in groups of densely connected regions. The MCODE heuristic screens for clique-like structures in a given PPI network. ClusterONE, SPICi and RNSC methods similarly find clusters satisfying high subgraph density/cohesiveness. CFinder and CMC enumerate clique structures, while DME enumerates all subgraphs exceeding a minimum density. The NeMo method aggregates nodes having many shared neighbors. The Nibble method utilizes the conductance objective function that is based on having many edges within clusters as opposed to going out of clusters. The random walk-based methods (RRW and AP) similarly find groups of densely connected subgraphs. Finally, graph-cut algorithms minimize the edge cut between clusters. In general, the common objective among these methods is to find dense network regions. These methods assume that a functional module corresponds to a strongly connected subgraph. However, clusters in PPIs are not always topologically dense. For instance, proteins in signaling pathways are often sparsely connected and may share important functions. Hence, such groups of proteins should also be presented in clustering results as they have significant biological implications. Techniques such as [89], VI-Cut and Tree-Sniping go beyond using dense structure to identify functional modules. However, these methods either cannot handle edge weights or fail to support high-dimensional GO annotations in a scalable manner.

3.5 Conclusions

In this chapter we have reviewed and classified some well known clustering techniques for PPI networks. Our approach has been to emphasize the unique characteristics of the network clustering problem in the context of PPI networks and discuss an array of techniques highlighting their strengths and limitations. This analysis is fundamental in developing effective computational solutions toward comprehending the organization and functioning of cells. Nevertheless, we do not claim completeness of this review and acknowledge any omissions.

References

1. A. Zhang, *Protein Interaction Networks: Computational Analysis* (Cambridge University Press, 2009)
2. S.S. Bhowmick, B.-S, Seah, Clustering and Summarizing Protein-Protein Interaction Networks: A Survey. IEEE Trans. Knowl. Data Eng. **28**(3), 638–658 (2016)
3. J. Ji, A. Zhang, et al., Functional module detection from protein-protein interaction networks, in *IEEE TKDE*, vol. 26, issue no. 2, 2014
4. F. Radicchi, C. Castellano, et al., Defining and identifying communities in networks. PNAS **101**(9) (2004)
5. F. Luo, Y. Yang et al., Modular organization of protein interaction networks. Bioinformatics **23**(2), 207–214 (2007)
6. M.P.H. Stumpf, T. Thorne et al., Estimating the size of the human interactome. PNAS **105**(19), 6959–6964 (2008)
7. M.J. Barber, Modularity and community detection in bipartite networks. Phys. Rev. **76**(6) (2007)
8. M.E.J. Newman, Modularity and community structure in networks. PNAS **103**(23) (2006)
9. J. Ruan, W. Zhang, An efficient spectral algorithm for network community discovery and its applications to biological and social networks, in *Proceedings of ICDM*, 2007, pp. 643–648
10. U. Brandes, D. Delling et al., On finding graph clusterings with maximum modularity. Graph-Theoretic Concepts in Computer Science, 2007, pp. 121–132
11. X. Xu, N. Yuruk, Z. Feng, T.A.J. Schweiger, Scan: a structural clustering algorithm for networks, in *In ACM SIGKDD*, 2007
12. H. Sun, J. Huang, et al., gskeletonclu: Density-based network clustering via structure-connected tree division or agglomeration, in *IEEE ICDM*, 2010
13. M. Newman, M. Girvan, Finding and evaluating community structure in networks. Phys. Rev. **69**(2) (2004)
14. J. Huang, H. Sun, et al., Shrink: a structural clustering algorithm for detecting hierarchical communities in networks, in *ACM CIKM*, 2010
15. A. Clauset, M.E.J. Newman, C. Moore, Finding community structure in very large networks. Phys. Rev. E **70**(6) (2004)
16. G. Karypis, V. Kumar, A fast and high quality multilevel scheme for partitioning irregular graphs. SIAM J. Sci. Comput. **20**, 359–392 (1998). Jan
17. D.A. Spielman, S.-H. Teng, A local clustering algorithm for massive graphs and its application to nearly-linear time graph partitioning, Sept 2008
18. Y. Zhou, H. Cheng, J.X. Yu, Clustering large attributed graphs: an efficient incremental approach, in *IEEE ICDM*, 2010
19. T. Nepusz, H. Yu, A. Paccanaro, Detecting overlapping protein complexes in protein-protein interaction networks. Nat. Methods **9**, 471–472 (2012)
20. C.G. Rivera, R. Vakil, J.S. Bader, NeMo: network module identification in cytoscape. BMC Bioinform. **11**(Suppl 1), S61 (2010). Jan
21. S. Asur, D. Ucar, S. Parthasarathy, An ensemble framework for clustering protein-protein interaction networks. Bioinformatics (Oxford, England) **23**, i29–40 (2007)
22. H.N. Chua, W.-K Sung, L. Wong, Exploiting indirect neighbours and topological weight to predict protein function from protein–protein interactions. Bioinformatics **22**(13) (2006)
23. S. Navlakha, J. White, N. Nagarajan, M. Pop, C. Kingsford, Finding biologically accurate clusterings in hierarchical tree decompositions using the variation of information. J. Comput. Biol. (J. Comput. Mol. Cell Biol.) **17**, 503–516 (2010). Mar
24. C. Kingsford, S. Navlakha, Exploring biological network dynamics with ensembles of graph partitions, in *Pacific Symposium on Biocomputing*, 2010, pp. 166–77
25. G.D. Bader, C.W.V. Hogue, An automated method for finding molecular complexes in large protein interaction networks. BMC Boinform. **4**, 2 (2003). Jan
26. A.C. Gavin, M. Bosche et al., Functional organization of the yeast proteome by systematic analysis of protein complexes. Nature **415**, 141–147 (2002)

27. M.C. Costanzo, M.E. Crawford, et al., YPD, PombePD and WormPD: model organism volumes of the BioKnowledge Library, an integrated resource for protein information. Nucleic Acids Res. **29**(1), 75–79 (2001)
28. A.H. Tong, B. Drees et al., A combined experimental and computational strategy to define protein interaction networks for peptide recognition modules. Science **295**, 321–324 (2001)
29. P. Uetz, L. Giot, G. Cagney, T.A. Mansfield, R.S. Judson, J.R. Knight, D. Lockshon, V. Narayan, M. Srinivasan, P. Pochart, A. Qureshi-Emili, Y. Li, B. Godwin, D. Conover, T. Kalbfleisch, G. Vijayadamodar, M. Yang, M. Johnston, S. Fields, J.M. Rothberg, A comprehensive analysis of protein-protein interactions in Saccharomyces cerevisiae. Nature **403**, 623–627 (2000). Feb
30. B.L. Drees, B. Sundin, et al., A protein interaction map for cell polarity development. PNAS **154**(3) (2001)
31. A.E. Mayes, L. Verdone, et al., Characterization of Sm-like proteins in yeast and their association with U6 snRNA. EMBO J. **18**(15) (1999)
32. T. Ito, T. Chiba, R. Ozawa, M. Yoshida, M. Hattori, Y. Sakaki, A comprehensive two-hybrid analysis to explore the yeast protein interactome. Proc. Natl. Acad. Sci. USA **98**, 4569–4574 (2001)
33. M. Altaf-Ul-Amin, Y. Shinbo, et al., Development and implementation of an algorithm for detection of protein complexes in large interaction networks. BMC Bioinform. **7**(1) (2006)
34. I. Xenarios, L. Salwinski et al., DIP, the Database of Interacting Proteins: a research tool for studying cellular networks of protein interactions. Nucleic Acids Res. **30**(1), 303–305 (2002)
35. M. Li, J.-E Chen, et al., Modifying the DPClus algorithm for identifying protein complexes based on new topological structures. BMC Bioinform. **9**(1) (2008)
36. A.D. King, N. Przulj, I. Jurisica, Protein complex prediction via cost-based clustering. Bioinformatics (Oxford, England) **20**, 3013–3020 (2004)
37. C. von Mering, R. Krause et al., Comparative assessment of largescale data sets of protein-protein interactions. Nature **417**, 399–403 (2002)
38. L. Giot, J.S. Bader et al., A protein interaction map of Drosophila melanogaster. Science **302**, 1727–1736 (2003)
39. S. Li, C.M. Armstrong et al., A map of the interactome network of the metazoan C.elegans. Science **303**, 540–543 (2004)
40. P. Pei, A. Zhang, A "seed-refine" algorithm for detecting protein complexes from protein interaction data. IEEE Trans. Nanobiosci. **6**(1), 43–50 (2007)
41. A.C. Gavin, P. Aloy et al., Proteome survey reveals modularity of the yeast cell machinery. Nature **440**, 431–436 (2006)
42. N.J. Krogan, G. Cagney, H. Yu, G. Zhong, X. Guo, A. Ignatchenko, J. Li, S. Pu, N. Datta, A.P. Tikuisis, T. Punna, J.M. Peregrín-Alvarez, M. Shales, X. Zhang, M. Davey, M.D. Robinson, A. Paccanaro, J.E. Bray, A. Sheung, B. Beattie, D.P. Richards, V. Canadien, A. Lalev, F. Mena, P. Wong, A. Starostine, M.M. Canete, J. Vlasblom, S. Wu, C. Orsi, S.R. Collins, S. Chandran, R. Haw, J.J. Rilstone, K. Gandi, N.J. Thompson, G. Musso, P. St, Onge, S. Ghanny, M.H.Y. Lam, G. Butland, A.M. Altaf-Ul, S. Kanaya, A. Shilatifard, E. O'Shea, J.S. Weissman, C.J. Ingles, T.R. Hughes, J. Parkinson, M. Gerstein, S.J. Wodak, A. Emili, J.F. Greenblatt, Global landscape of protein complexes in the yeast Saccharomyces cerevisiae. Nature **440**, 637–643 (2006)
43. S.R. Collins, P. Kemmeren et al., Toward a comprehensive atlas of the physical interactome of Saccharomyces cerevisiae. Mol. Cell Proteomics **6**(3), 439–450 (2007)
44. P. Jiang, M. Singh, SPICi: a fast clustering algorithm for large biological networks. Bioinformatics (Oxford, England) **26**, 1105–1111 (2010)
45. L.J. Jensen, M. Kuhn et al., STRING 8a global view on proteins and their functional interactions in 630 organisms. Nucleic Acids Res. **37**, D412–D416 (2009)
46. C. Huttenhower, E.M. Haley, et al., Exploring the human genome with functional maps. Genome Res. **19**(6) (2009)
47. A.J. Enright, S. Van Dongen, C.A. Ouzounis, An efficient algorithm for large-scale detection of protein families. Nucleic Acids Res. **30**, 1575–1584 (2002). Apr

48. V. Satuluri, S. Parthasarathy, D. Ucar, Markov clustering of protein interaction networks with improved balance and scalability, in *Proceedings of the First ACM International Conference on Bioinformatics and Computational Biology - BCB '10*, (ACM Press, New York, New York, USA, 2010), p. 247
49. S. Razick, G. Magklaras, I.M. Donaldson, iRefIndex: a consolidated protein interaction database with provenance. BMC Bioinform. **9** (2008)
50. Y.-K. Shih, S. Parthasarathy, Identifying functional modules in interaction networks through overlapping Markov clustering. Bioinformatics (Oxford, England) **28**, i473–i479 (2012)
51. L. Kiemer, S. Costa et al., WI-PHI: a weighted yeast interactome enriched for direct physical interactions. Proteomics **7**, 932–943 (2007)
52. J.B. Pereira-Leal, A.J. Enright, C.A. Ouzounis, Detection of functional modules from protein interaction networks. PROTEINS: Struct. Funct. Bioinform. **54**(1), 49–57 (2004)
53. Y.-R Cho, W. Hwang, M. Ramanathan, A. Zhang, Semantic integration to identify overlapping functional modules in protein interaction networks. BMC Bioinform. **8**(1) (2007)
54. Y. Cho, L. Shi, A. Zhang, Functional module detection by functional flow pattern mining in protein interaction networks. BMC Bioinform. **9** (2008)
55. X. Lei, X. Huang, L. Shi, A. Zhang, Clustering PPI data based on improved functional-flow model through quantum-behaved PSO. Int. J. Data Mining Bioinform. **6**(1), 42–60 (2012)
56. V. Spirin, L.A. Mirny, Protein complexes and functional modules in molecular networks. Proc Natl Acad Sci USA **100**, 12123–12128 (2003). Oct
57. B. Adamcsek, G. Palla, et al., CFinder: locating cliques and overlapping modules in biological networks. Bioinformatics **22**(8) (2006)
58. S. Zhang, X. Ning, X.-S. Zhang, Identification of functional modules in a PPI network by clique percolation clustering. Comput. Biol. Chem. **30**(6), 445–451 (2006)
59. G. Cui, Y. Chen, et al., An algorithm for finding functional modules and protein complexes in protein-protein interaction networks. J. Biomed. Biotechnol. (2008)
60. A. Ruepp, A. Zollner et al., The FunCat, a functional annotation scheme for systematic classification of proteins from whole genomes. Nucleic Acids Res. **32**(18), 5539–5545 (2004)
61. G. Liu, L. Wong, H.N. Chua, Complex discovery from weighted PPI networks. Bioinformatics (Oxford, England) **25**, 1891–1897 (2009)
62. Y. Ho, A. Gruhler et al., Systematic identification of protein complexes in Saccharomyces cerevisiae by mass spectrometry. Nature **415**, 180–183 (2002)
63. P. Aloy, B. Bottcher et al., Structure-based assembly of protein complexes in yeast. Science **303**, 2026–2029 (2004)
64. E. Georgii, S. Dietmann, T. Uno, P. Pagel, K. Tsuda, Enumeration of condition-dependent dense modules in protein interaction networks. Bioinformatics (Oxford, England) **25**, 933–940 (2009)
65. U. Guldener et al., MPact: the MIPS protein interaction resource on yeast. Nucleic Acids Res. **34**, D436–D441 (2006)
66. S. Kerrien, Y. Alam-Faruque, B. Aranda, I. Bancarz, a. Bridge, C. Derow, E. Dimmer, M. Feuermann, A. Friedrichsen, R. Huntley, C. Kohler, J. Khadake, C. Leroy, a. Liban, C. Lieftink, L. Montecchi-Palazzi, S. Orchard, J. Risse, K. Robbe, B. Roechert, D. Thorneycroft, Y. Zhang, R. Apweiler, H. Hermjakob, IntAct–open source resource for molecular interaction data. Nucleic Acids Res. **35**, D561–D565 (2007)
67. B.J. Frey, D. Dueck, Clustering by passing messages between data points. Science (New York, NY) **315**, 972–976 (2007). Feb
68. K. Macropol, T. Can, A.K. Singh, RRW: repeated random walks on genome-scale protein networks for local cluster discovery. BMC Bioinform. **10**, 283 (2009). Jan
69. J. Chen, B. Yuan, Detecting functional modules in the yeast protein–protein interaction network. Bioinformatics **22**(18) (2006)
70. J.M. Cherry, C. Adler, et al., SGD: Saccharomyces genome database. Nucleic Acids Res. **26**(1) (1998)
71. D. Dotan-Cohen, A.A. Melkman, S. Kasif, Hierarchical tree snipping: clustering guided by prior knowledge. Bioinformatics (Oxford, England) **23**, 3335–3342 (2007)

72. M. Mete, F. Tang, X. Xu, N. Yuruk, A structural approach for finding functional modules from large biological networks. BMC Bioinform. **9** (2008)
73. M. Jayapandian, A. Chapman et al., Michigan Molecular Interactions (MiMI): putting the jigsaw puzzle together. Nucleic Acids Res. **35**, D566–D571 (2006)
74. D. Greene, G. Cagney, N. Krogan, P. Cunningham, Ensemble non-negative matrix factorization methods for clustering protein-protein interactions. Bioinformatics **24**(15), 1722–1728 (2008)
75. E. Segal, H. Wang, D. Koller, Discovering molecular pathways from protein interaction and gene expression data. Bioinformatics **19** (2003)
76. A.P. Gasch, P.T. Spellman et al., Genomic expression program in the response of yeast cells to environmental changes. Mol. Biol. Cell **11** (2000)
77. P.T. Spellman, G. Sherlock, et al., Comprehensive identification of cell cycle-regulated genes of the yeast *Saccharomyces cerevisiae* by microarray hybridization. Mol. Biol. Cell **9**(12) (1998)
78. H. Lu, B. Shi et al., Integrated analysis of multiple data sources reveals modular structure of biological networks. Biochem. Biophys. Res. Commun. **345**(1), 302–309 (2006)
79. W.K. Huh, J.V. Falvo et al., Global analysis of protein localization in budding yeast. Nature **425**, 686–691 (2003)
80. J.M. Stuart, E. Segal, D. Koller, S.K. Kim, A Gene-coexpression network for global discovery of conserved genetic modules. Science **302** (2003)
81. I.A. Maraziotis, K. Dimitrakopoulou, A. Bezerianos, Growing functional modules from a seed protein via integration of protein interaction and gene expression data. BMC Bioinform. **8**(1) (2007)
82. A. Patil, H. Nakamura, Filtering high-throughput protein-protein interaction data using a combination of genomic features. BMC Bioinform. **6**(100) (2005)
83. H. Zheng, H. Wang, D.H. Glass, Integration of genomic data for inferring protein complexes from global protein–protein interaction networks. IEEE Trans. Syst. Man Cybern. Part B: Cybern. **38**(1) (2008)
84. L.J. Lu, Y. Xia, et al., Assessing the limits of genomic data integration for predicting protein networks. Genome Res. **15**(7) (2005)
85. T.R. Hughes, M.J. Marton, et al., Functional discovery via a compendium of expression profiles. Cell **102**(1) (2000)
86. R.J. Cho, M.J. Campbell et al., A genome-wide transcriptional analysis of the mitotic cell cycle. Mol Cell. **2**(1), 65–73 (1998)
87. I. Ulitsky, R. Shamir, Identifying functional modules using expression profiles and confidence-scored protein interactions. Bioinformatics **25**(9), 1158–1164 (2009)
88. A.P. Gasch, M. Huang, et al., Genomic expression responses to DNA-damaging agents and the regulatory role of the yeast ATR homolog Mec1p. Mol. Biol. Cell **12**(10) (2001)
89. L. Shi, X. Lei, A. Zhang, Detecting protein complexes with semi-supervised learning in protein interaction networks. Proteome Sci. **9** (2011)
90. H. Wang, W. Wang, J. Yang, P. Yu, Clustering by pattern similarity in large data sets, in *ACM SIGMOD*, 2002
91. J. Sun, B. Feng, W.B. Xu, Particle swarm optimization with particles having quantum behavior, in *IEEE Proceedings of Congress on Evolutionary Computation*, 2004
92. G. Palla, I. Derényi, I. Farkas, T. Vicsek, Uncovering the overlapping community structure of complex networks in nature and society. Nature **435**, 814–818 (2005). June
93. K. Voevodski, S.-H Teng, Y. Xia, Finding local communities in protein networks. BMC Bioinform. **10**(1) (2009)
94. J. Vlasblom, S.J. Wodak, Markov clustering versus affinity propagation for the partitioning of protein interaction graphs. BMC Bioinform. **10**, 99 (2009). Jan
95. M. Girvan, M.E.J. Newman, Community structure in social and biological networks. PNAS **99**(12) (2002)
96. I. Ulitsky, R. Shamir, Identification of functional modules using network topology and high-throughput data. BMC Syst. Biol. **8**(1) (2007)
97. S. Brohee, J. van Helden, Evaluation of clustering algorithms for protein-protein interaction networks. BMC Bioinform. **7**(1) (2006)

Chapter 4
Functional Summarization

In the preceding chapter, we have discussed an array of techniques for PPI network clustering. In this chapter, we explore recent work in PPI *network summarization*, a problem that is closely related to clustering. Specifically, we focus on constructing higher level *functional summary* that summarizes the underlying PPI network to obtain a concise, interpretable representation of the network. We begin by motivating the need for PPI network summarization. Then, we highlight the limitations of aforementioned clustering techniques to address the summarization problem effectively. Finally, we present a recently proposed functional summarization technique called FUSE [1] in addressing the information overload issue of analyzing large scale PPI networks. We evaluate the performance of FUSE on several real-world PPIs. We also compare FUSE to state-of-the-art graph clustering methods with GO term enrichment by constructing the biological process landscape of the PPIs. Our experimental results demonstrate that FUSE is highly effective in constructing higher order functional maps with superior accuracy and representativeness compared to these state-of-the-art graph clustering methods. Using Alzheimer's Disease network as our case study, we further demonstrate the ability of FUSE to quickly summarize the network and identify many different processes and complexes that regulate it. We analyze the topological features of the functional landscape of human PPI that leads us to the identification of *functional hubs* (clusters of proteins that act as hubs).

4.1 Motivation

Recall that with advances in high throughput experimental biology, the number of large scale protein interaction networks (PPI) have grown rapidly. At the same time, collaborative efforts to annotate proteins and genes using GO annotations has generated detailed attributes that describe these entities. Knowledge-bases with GO annotations, such as UniprotKB [2], provide a wealth of annotation data at different levels

S.S. Bhowmick and B.-S. Seah, *Summarizing Biological Networks*,
Computational Biology 24, DOI 10.1007/978-3-319-54621-6_4

of specificity. Recall from Chap. 2 that GO provides standardized annotations that describe various attributes of a gene or protein, including localization attributes, molecular function, and the biological processes it participates in. As proteins may involve in multiple roles and functions, GO attributes associated with a protein or a gene can be high-dimensional.

As highlighted in Chap. 1, the amount of information contained within large biological networks can often overwhelm researchers, making systems level analysis of PPIs a daunting task. As majority of function annotation and high throughput or curated interaction data are encoded at protein or gene level, higher-order abstraction maps such as complex-complex or process-process functional landscapes, are often unavailable. However, availability of such information is invaluable as it not only allows one to ask questions about the relationships among high-level modules, such as processes and complexes, but also allows one to visualize higher order patterns from a bird's eye perspective.

For instance, consider the Alzheimer's Disease (AD) related PPI in *IntAct* [3]. An AD interaction network can be studied at different levels of organization, from broad-level process-process interactions to in-depth complex-complex interactions. Such maps would reveal higher-level patterns that otherwise would have been invisible. The objective here is not to study a process associated with AD in isolation, but instead focus on the interplay of related processes in tandem to identify the causative mechanisms of AD. For example, one might ask the following questions: How do signaling pathways implicated for AD associate with one another? How do proteins related to transportation play a role in AD, and how are they associated with bioenergetics? A bird's-eye view of the functional landscape of AD network may provide answers to these questions. An example is shown in Fig. 4.1 (detailed in Sect. 4.8). Observe that the associations between signaling pathways (*A28*, *A14*, *A18*, *A21*, and *A16*) are depicted in the summary. It is worth mentioning that it is extremely difficult to answer the aforementioned questions by simply looking at a large PPI containing a large number of proteins and interactions. This problem is further exacerbated by the high-dimensional nature of PPI; each protein may have hundreds of annotation attributes. *It is therefore crucial to have some form of summarization that maps higher-order information of the underlying* PPI. Fortunately, the modular nature of biological networks–either structurally or attribute wise–lends itself to the possibility of building such a summary.

4.2 Limitations of PPI Clustering Techniques

Although tools to abstract high-level and functional information from gene lists have been proven to be key to analyzing high throughput data [4], similar tools that automatically abstract and *summarize* PPIs at multiple resolutions to provide high level views of functional landscape of PPIs are still lacking. At first glance, it may seem that state-of-the-art graph clustering techniques [5–9] that are discussed in the preceding chapter can be used for generating high quality summaries of PPIs

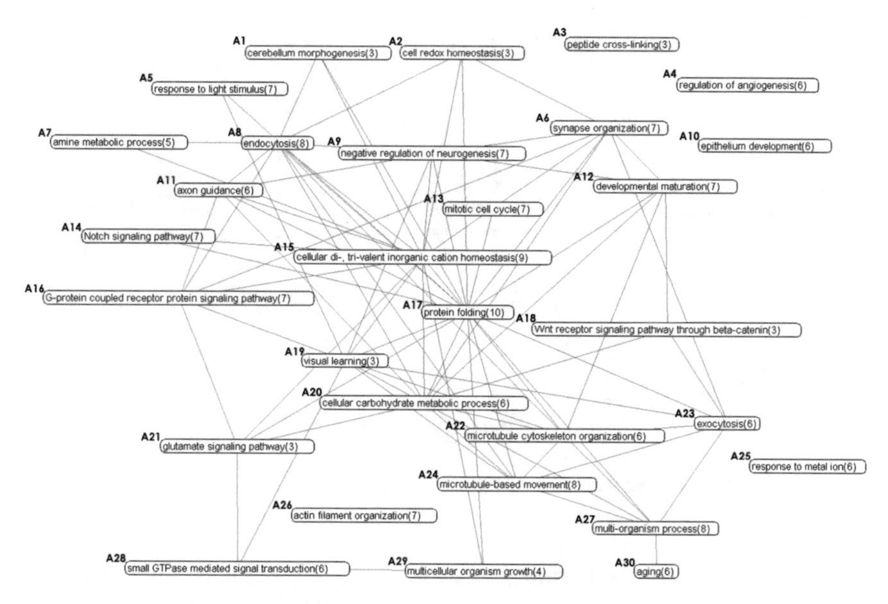

Fig. 4.1 Functional summary (FSG) of the AD network for $k = 30$ (cluster size indicated in brackets)

as these techniques have been successful in identification of novel protein function and protein complexes. Intuitively, a biological network can be decomposed into modules–groups of vertices sharing a common function–that are then collapsed into a representative node to form a summary graph of the underlying network. Depending on the granularity of the decomposition, summaries of various level of detail can be formed. Despite the benefits of graph clustering in creating summary of a PPI network, these techniques suffer from the following key weaknesses that make them less suitable for building high quality higher order *functional summaries* of PPIs.

Firstly, several existing graph clustering approaches [5–7, 10] overwhelmingly emphasize structure cohesiveness over attribute coherence. However, in practical applications of PPI summarization, attribute coherence is key to forming meaningful, interpretable modules. In PPI, groups of proteins (vertices) that share a common vertex property can form a meaningful cluster that represents a particular biological function. Otherwise, clusters with inconsistent vertex properties, even if structurally well-connected, may not simply summarize into one functionally interpretable cluster. Secondly, majority of existing graph clustering techniques form non-overlapping partitions [5, 7, 10]. Consequently they cannot be used to generate high-quality summary as "interactors" in biological processes and pathways are likely to overlap [11]. Thirdly, these techniques typically focus on identifying dense subgraphs from a graph. However, higher-level clusters in PPIs are not always structurally dense. Proteins in signaling pathways, for instance, are structurally loose, but share important functions. Such groups of proteins often have significant biological implications despite their loose structure, and should be present in any summary of the underlying

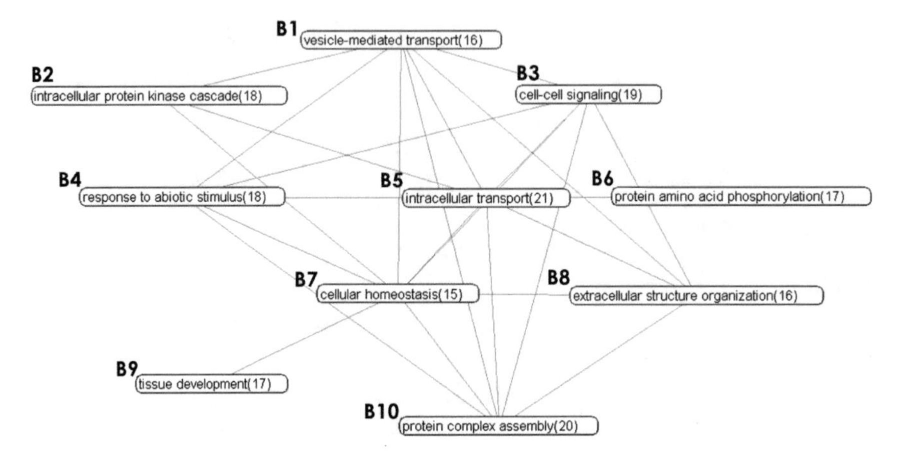

Fig. 4.2 FSG of the AD network ($k = 10$)

network. Finally, because the annotations that describe proteins and their functions are high-dimensional, finding the right choice of attribute coherent groupings is combinatorial and non-trivial (for example, Fig. 4.2).

Figure 4.3 contains examples of both optimal and sub-optimal clustering-based summarization of biological networks. In Fig. 4.3a, an optimal summarization decomposes the graph into clusters A and B, both of which have consistent attributes and cohesive structure. Although the underlying graph is a clique, choosing a cluster that encompasses all vertices–as shown in Fig. 4.3b–would be sub-optimal, because vertex attributes within the cluster would then be inconsistent. Consequently, one could not extract a common, biologically meaningful function that represents the

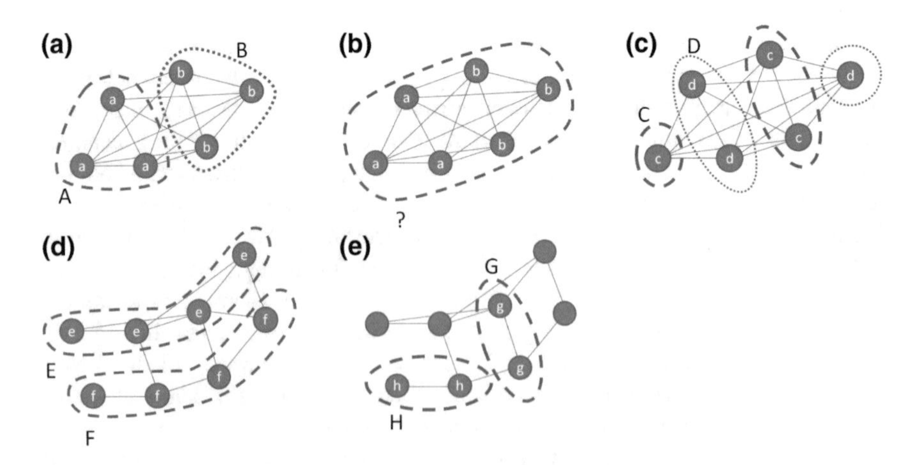

Fig. 4.3 Structure and attribute considerations in network summarization

cluster. Figure 4.3c is also less optimal compared to Fig. 4.3a because despite having attribute consistent clusters, intra-cluster cohesiveness of the vertices are weak.

Let us consider another scenario. Figure 4.3d shows a graph that is partitioned in clusters E and F. In the absence of dense structure and coherent attributes, it may be partitioned based on its attributes. Finally, Fig. 4.3e shows an example of poor summarization, as despite having consistent attributes, the clusterings have inadequate coverage. This makes the subsequent summary less representative of the underlying network.

4.3 Overview

We present a data-driven algorithm called FUSE (**Fu**nctional **S**ummary g**E**nerator) that addresses the aforementioned challenges (Sects. 4.5 and 4.6). Given a PPI network, it generates a *k-node functional summary graph* (FSG) that best represents the higher-order abstraction of the PPI network by simultaneously evaluating interaction and annotation data. We argue that a "good" functional summary of a network is not merely a graph of all function-function relationships, but a graph that *reduces* details of the original PPI network to form a subset of interconnected *functional clusters*. A *functional cluster* represents a subnetwork of proteins that shares a common function. In particular, the functional summary graph must simultaneously satisfy the following requirements: (a) the summary is at a *specific level* (k nodes) of detail, (b) the summary is representative of the original network, and (c) redundancies are minimized. Specifically, FUSE exploits *Minimum Description Length* principle [12] to generate the "best" summary by maximizing *information gain* while satisfying the level of details constraint. Figures 4.1 and 4.2 depict a 30-node and a 10-node FSGs of the AD network, respectively, generated by FUSE.

The goal of FUSE is not only to generate a higher level functional summary that is representative of the underlying PPI network, but also to generate a k-node functional map whose visual complexity (determined by k) permits user analysis. With 30,000 or more terms in the GO, interaction network of 30,000 functional modules will not be a useful summary, as it is just as daunting as the original PPI network, if not more. FUSE addresses this challenge by enabling generation of summaries that are small and understandable.

4.4 Related Work

Functional landscape of an underlying protein interaction network has been explored in [13]. The approach the authors used, however, rely on manual short listing of 229 biological processes for analysis. While this approach makes visualization permissible, it neither scales with the growing number of annotations, nor does it fully utilize the availability of large number of annotations. Additionally, the processes that are relevant depend on the context of the network.

Traditionally, functions pertaining to a list of genes are extracted through functional enrichment techniques, which identify statistically over-represented functions within a set of proteins or genes [4]. Such approaches are designed to identify enriched functions that describe the dataset as whole. In contrast to FUSE, they do not utilize the interaction data encoded within PPI, which is key to understanding how the processes and complexes cooperate to govern a particular function.

As discussed in the preceding chapter, network clustering methods, on the other hand, identify functional clusters based on the underlying assumption that the topology of interacting proteins can be mined to identify protein clusters [5–7, 10]. Cluster function can then be inferred and annotated by finding enriched annotations within the cluster. While such methods have been proven effective for identification of complexes, they are less suitable for identifying higher level functional clusters, such as biological processes and pathways, where interactors within them are likely to overlap [11, 14]. Interactions within a process are also not necessarily cohesive. CFinder [15] locates overlapping communities based on structure of the network, but ignores the wealth of functional knowledge already encoded in GO annotation data. While most graph clustering techniques rely solely on network topology, several recent techniques utilize annotation information when clustering the networks [8, 9, 16, 17]. However, these techniques form non-overlapping partitions. Additionally, with the growing amount of annotation data, the attribute space of the nodes in an interaction network is high dimensional as a single protein may be linked to hundreds of annotations. However, these state-of-the-art approaches are not designed for clustering high-dimensional attributes of GO annotated interaction networks. For instance, in [8], a "semantic" distance function is used to measure semantic similarities between nodes with multiple MIPS complex annotations. The *curse of dimensionality* limits the applicability of such an approach on GO annotations [18]. Note that existing subspace clustering approaches that allow overlapping subspace clusters typically produce a huge number of clusters that are difficult to interpret [19].

Lastly, the high dependency on interaction topology makes graph clustering ineffective for many context-specific networks. Although there are many networks associated with diseases, there are few, if any, with complete interaction knowledge available. The high probability of false positive interactions may also occur. This hampers accurate identification of cohesive clusters.

Recently, there has been increasing research by the data mining community towards generating effective summary of large graph-structured data [16, 20–22]. For instance, Tian et al. [16] focus on grouping nodes at different resolutions in a large network based on user-selected node attributes and relationships. VoG [20] finds a set of possibly overlapping subgraphs based on most important local structures such as cliques, bipartite cores, stars, and chains. However, these techniques cannot be leveraged to summarize PPI networks because of the following key reasons. First, approaches such as [16, 21, 22] focus on non-overlapping groups instead of overlapping ones which are desirable in PPI networks. Second, similar to existing graph clustering techniques, VoG [20] and GraSS [22] do not leverage annotations associated with nodes in PPI networks and hence may result in summaries that are not biological meaningful for reasons discussed earlier. Furthermore, these techniques do not attempt to construct summaries that maximally cover the input network while minimizing redundancy among the summary subgraphs.

4.5 The Functional Summarization Problem

In this section, we formally introduce the functional summarization problem. We begin by defining some terminology that we shall be using in the sequel. A list of key notations used in this chapter is given in Table 4.1.

Recall that a PPI network $G = (V, E)$ contains a set of vertices V, representing proteins, and a set of edges E, representing interactions. An edge has a positive real weight ω that represents its interaction strength. Given a GO directed acyclic graph (DAG), denoted as D, the ordered set $\Delta = \langle a_1, a_2, \ldots, a_n \rangle$ is a topological sort of D, where a_i represents a single GO term. The *term association vector* of $v \in V$, denoted by Δ_v, is defined as $\Delta_v = \langle a_1(v), a_2(v), \ldots, a_n(v) \rangle, a_i(v) \in \{0, 1\}$, such that $a_i(v) = 1$ if and only if the term a_i or its descendants are associated with protein v. Otherwise, $a_i(v) = 0$. Note that Δ_v indicates GO terms that are associated with v.

4.5.1 Functional Summary of PPI

Given a PPI network $G = (V, E)$, a *functional summary graph* (FSG) is an undirected graph $\Theta_G = (S, F)$ that models the set of higher-order *functional clusters* S and their

Table 4.1 Notations

Symbols	Description
G	Input PPI graph
$\Theta_G = (S, F)$	Functional summary graph where S and F are sets of nodes and interactions, respectively
ω	Edge weight
Δ	Set of GO terms
S_Δ	Set of functional clusters induced from Δ
$C(u)$	Functional cluster representing the function u
$\phi^{C(u)}$	Structural information content of cluster $C(u)$
$c^{C(u)}$	Size deviation cost
k	Summary complexity parameter
b	Information budget parameter
d	Redundancy penalizing parameter
β	Significance cut-off parameter

interactions F that underlie the PPI network. A *functional cluster* is a subgraph of G that shares a particular function/role based on the structure and attribute properties of the subgraph and its constituent proteins. Functional clusters may include complexes, processes, and signaling pathways. A pair of functional clusters may be connected by a web of protein interactions. If the number of interactions are significantly large, then we say that the pair of clusters are *associated.* An FSG Θ_G thus captures higher order modules that comprise the PPI and their interconnections. We now define these concepts formally.

Definition 4.1 (*Functional Cluster*) Let $V(a_i) \subseteq V$ denote the set of vertices in G such that $v \in V(a_i)$ if and only if $\Delta_v[a_i(v)] = 1$. The *functional cluster* of $a_i \in \Delta$, denoted by $C(a_i) \subseteq G$, is the subgraph of G that is induced by $V(a_i)$.

Note that $V(a_i)$ represents the set of vertices of G that are associated with term $a_i \in \Delta$. We treat $C(a_i)$ as a vertex as well. We may also call a functional cluster as *functional subgraph* when we wish to emphasize the fact that it is a graph. Figure 4.4b shows a subset of the possible functional clusters of the PPI in Fig. 4.4a. Every node in

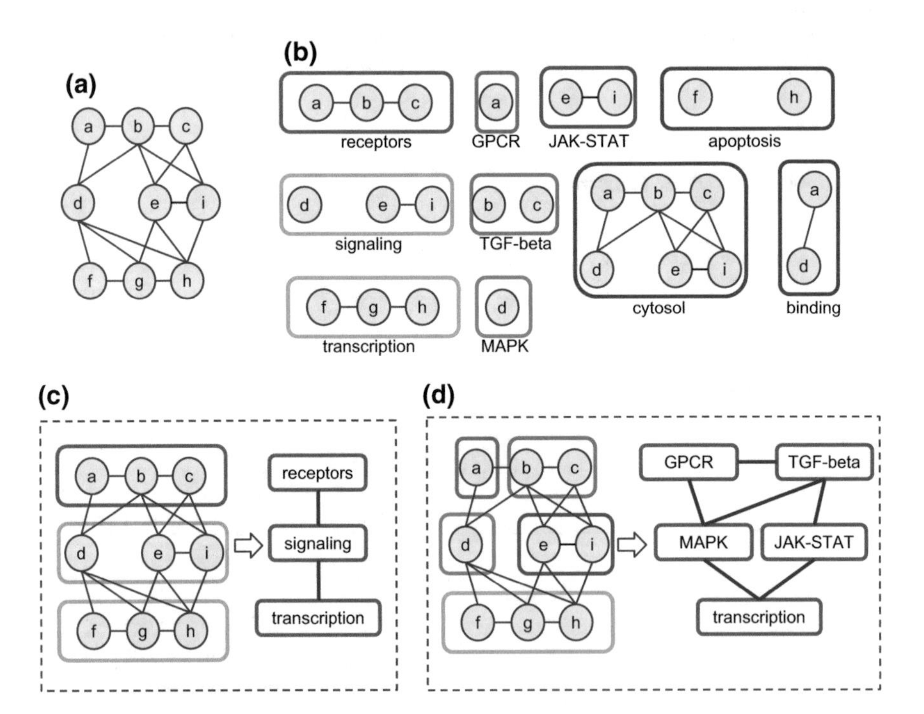

Fig. 4.4 **a** A toy example of PPI network. **b** A set of *functional clusters* of the network in (**a**). **c** Suppose a 3-node summary is required ($k = 3$). FUSE explores the functional clusters of the PPI network to identify the 3-node functional summary that best partition and represent the underlying network. This functional summary graph (FSG) depicts the functional landscape of the PPI network in 3 nodes. **d** A 5-node partition ($k = 5$) and its corresponding FSG

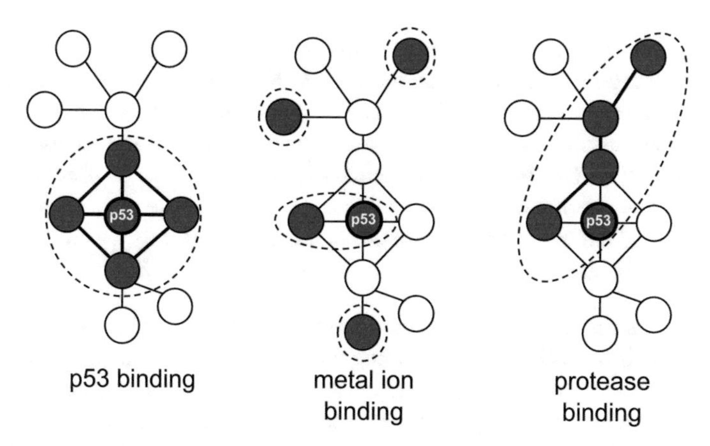

Fig. 4.5 Functional clusters associated with the p53 protein

a cluster must share a particular function or attribute. For instance, nodes in functional cluster `cytosol` share the `cytosol` term.

Given that a protein could be annotated with multiple GO terms, there are a multitude of ways to form functional clusters. To illustrate this, let us restrict ourselves to several GO terms associated with the `p53` protein, namely `'p53 binding'`, `'metal ion binding'` and `'protease binding'` terms. Figure 4.5 shows a toy network of functional clusters formed using the `'p53 binding'`, `'metal ion binding'` and `'protease binding'` terms, respectively. A `'p53 binding'` functional cluster, for example, is constructed by first taking all proteins sharing the `'p53 binding'` term (indicated in the figure as shaded nodes). Following that, the subgraph induced by these proteins forms the `'p53 binding'` functional cluster.

The three functional clusters represent several alternative ways which proteins can be grouped together based on their shared function. Recall from the previous chapter, a good cluster exhibits significant clustering properties. For instance, proteins in a cluster should be densely interacting. Using the same intuition, we assess the clustering property of the three functional clusters using subgraph density. The `'p53 binding'` subgraph is most suitable as a cluster, given that the induced subgraph formed using proteins sharing this function has the highest subgraph density. Conversely, functional cluster formed using `'metal ion binding'` has the lowest subgraph density. This illustration demonstrates that although a gene or protein could have multiple GO term annotations, not all of these terms are useful for forming a functional cluster. It is therefore important to identify and distinguish terms that are suitable for forming functional clusters from those which are not. We shall see later in the next section how a suitable set of functional clusters can be selected to represent the entire PPI network.

Definition 4.2 (*Functional Summary Graph (FSG)*) A *functional summary graph* of the underlying protein interaction network $G = (V, E)$, Θ_G, is defined as $\Theta_G = (S, F, P_i, \alpha)$, where S is a set of functional clusters and F is a set of edges that links the functional clusters. Let oc_{uv} be the number of interactions connecting proteins in $C(u)$ and $C(v)$. Let P_i be the probability density function of observing o_{uv} or more number of interactions between $C(u)$ and $C(v)$. Let β be a significance cut-off parameter (user-defined). Then, $(C(u), C(v)) \in F$ if and only if $P_i(X > oc_{uv}) \leq 2\beta/|S|^2$. The bijection $\alpha : 1, 2, \ldots, m \leftrightarrow S$ is an ordering of S.

Observe that the aforementioned definition of functional summary includes additional constructs and rules for determining whether two functional clusters are associated. We elaborate on this further. Given a PPI $G = (V, E)$, the expected probability of observing an interaction between two randomly drawn protein pair is given by $p_i = \frac{2|E|}{|V|(|V|-1)}$. Let $(C(u), C(v))$ be a functional cluster pair such that members of both clusters were randomly drawn from V. If proteins v_1 and v_2 are randomly drawn from $C(u)$ and $C(v)$, respectively, then the expected probability of observing a positive interaction between them would also be p_i. Let $n = |C(u)||C(v)|$. Based on the independent and identically distributed variable (*iid*) assumption, we model the probability of observing oc (the number of interactions between $C(u)$ and $C(v)$) as the probability of observing oc positive interactions after n *iid* trials, representing n pairwise interaction trials between proteins in $C(u)$ and $C(v)$. Hence, the probability of oc or more positive interactions between $C(u)$ and $C(v)$ can be modeled using a binomial distribution:

$$P_i(X > oc) = \sum_{i=oc}^{n} \binom{n}{i} {p_i}^i (1 - p_i)^{n-i} \tag{4.1}$$

This $p - value$ is used to assess the *association significance* between a pair of functional clusters. Given a set containing k clusters, association significance between $\frac{1}{2}k(k-1)$ pairs of clusters would have to be tested. To this end, we applied Bonferroni correction to account for multiple testing. Given the *significance cut-off* β, a pair of functional clusters is *significantly associated* if

$$P_i(X > oc) \leq 2\beta/k(k-1) \approx 2\beta/k^2 \tag{4.2}$$

Observe that although we have adopted a simple model to assess cluster-cluster association, the aforementioned definition is general enough to encompass more sophisticated association models.

Example 4.1 Figure 4.4d shows an FSG consisting of five functional clusters. Any edge between two functional clusters exists when $P_i(X > oc_{uv}) \leq 2\beta/|S|^2$, implying that more edges connect proteins between the functional clusters than expected in random.

4.5.2 Problem Statement

The functional summarization problem is the problem of finding Θ_G that best represents the underlying PPI subject to a *summary complexity constraint*. To model this problem, we propose a profit maximization model that aims to find $\Theta_G = (S, F, P_i, \alpha)$ by maximizing information profit under a budget constraint. Every protein $i \in V$ is assigned a non-negative *information budget* b, which represents the information it contains. Let S_Δ be the set of functional clusters induced from Δ. Every functional cluster $C(u) \in S_\Delta$ is assigned a non-negative *structural information value* $\psi^{C(u)}$ (to be defined later), which represents the amount of structural information contained within the functional subgraph. When a functional cluster $C(u)$ is added to the summary, for every protein $i \in V(u)$, a portion of b is taken out and added to summary information gain. This represents new information added to the summary. The amount to take depends on $\psi^{C(u)}$.

Imposing information budget b limits the amount of information a protein can provide. A parameter $0 \le d \le 10$ is also introduced to penalize redundancy. By doing so, repeated representation of a protein i yields reduced information gain, modeling diminishing returns. Based on this profit model, we construct the set of functional clusters that maximizes profit while satisfying the constraints.

Definition 4.3 (*Functional Summarization Problem*) Let K_i be a set of functional clusters such that $C(u) \in K_i$ if and only if $i \in C(u)$. For every $C(u) \in S_\Delta$, let $\psi^{C(u)}$ be the structural information value of $C(u)$. Given a PPI network $G = (V, E)$ and user-defined parameters b, d and k, the *functional summarization problem* constructs a k-cluster FSG $\Theta_G = (S, F, P_i, \alpha)$ that satisfies the following optimization problem:

$$\begin{aligned}
&\text{maximize} \sum_{i \in V} \sum_{j=1}^{|S|} p(i, j) \\
&\text{where} \\
&b(i, m) = \begin{cases} \frac{d}{10}(b(i, m-1) - p(i, m-1)) & \text{if } m > 1, \\ & \alpha_S(m-1) \in K_i \\ b(i, m-1) & \text{if } m > 1, \\ & \alpha_S(m-1) \notin K_i \\ b & \text{if } m = 1 \end{cases} \\
&\text{and} \\
&p(i, m) = \begin{cases} \psi^{\alpha_S(m)} \text{ if } b(i, m) \ge \psi^{\alpha_S(m)} \text{ and } \alpha_S(m) \in K_i \\ b(i, m) \text{ if } b(i, m) < \psi^{\alpha_S(m)} \text{ and } \alpha_S(m) \in K_i \\ 0 \qquad \alpha_S(m) \notin K_i \end{cases} \\
&\text{subject to} \\
&|S| = k \\
&S \subset S_\Delta
\end{aligned} \tag{4.3}$$

Here, $p(i, j)$ serves as a store of profit obtained each time protein i is selected in one of the cluster at the j-th iteration. The map $\alpha_S(m)$ serves as an index set to assign the m-th iteration taken to its associated subgraph K of iteration m. The function $b(i, m)$, the remaining budgeted profit that can be taken per protein, and it can be derived recursively from its preceding $b(m-1)$. We elaborate on how the *structural information value* $\psi^{C(u)}$ is assigned. A functional cluster $C(u)$ and its protein constituents share a common function u. Thus, vertices in the subgraph are considered homogeneous attribute wise. However, it does not imply that the functional subgraph is structurally cohesive (dense). Proteins having common function u may still be weakly interacting. This may be due to the fact that u itself may indicate a general function (e.g., `'protein binding'`) which is a common attribute to a large number of proteins that do not interact with each other. We argue that structurally cohesive functional clusters contain more information than those which are loosely interconnected. The argument for this is that clusters that have higher than expected cohesiveness will have higher information content because of the lower probability of observing a random cluster having the same cohesiveness. However, we make the following exception – a functional cluster with lower than expected cohesiveness is not deemed structurally informative.

Since the optimization problem must choose among a set of functional clusters, we are not concerned about the actual p-value of observing a subgraph having such interaction density. Instead, we only need a measure that allows us to compute the relative ranking of the functional clusters by their information content. Such simplification leads to much greater computation efficiency. We define the *structural information value* of a functional cluster $C(u)$ as follows.

Definition 4.4 (*Structural Information Value*) Let ω_{ij} be the edge weight of $(i, j) \in E$. The *structural information value* of a functional cluster $C(u)$, denoted by $\psi^{C(u)}$, as $\psi^{C(u)} = \rho^{C(u)}$ where

$$\rho^{C(u)} = \frac{\sum_{i,j \in C(u)} E_{ij}}{|C(u)|} \tag{4.4}$$

At first glance, it may seem that the structural information value should be defined as $\psi^{C(u)} = \rho^{C(u)} - \rho^{\text{random}}$ where ρ^{random} is the *expected structural density* of a random cluster. However, we ignore ρ^{random} for the following reason. $\rho^{C(u)}$ is the *ratio association* [23] score of $C(u)$, a standard graph clustering objective we adopt that indicates the structural density of $C(u)$. In scale-free and Erdős–Rényi graphs, the self-information $-\log P(\psi^{C(u)})$ is a positive non-decreasing function of $\psi^{C(u)}$ for $\psi^{C(u)} > 0$. Hence, $\psi^{C(u)}$ can be used to compare the self-information between two functional clusters without having to determine the probability density function of the interaction distribution of a subgraph. Given $a_i, a_j \in \Delta$, $C(a_i)$ is deemed more informative than $C(a_j)$ if and only if $\psi^{C(a_j)} > \psi^{C(a_i)}$ and $\psi^{C(a_j)} > 0$. If both $\psi^{C(a_j)}$ and $\psi^{C(a_i)}$ are negative, it does not matter whether one is more informative than the other, since both have structural density less than that of random networks. As such, for symmetry, we also deem that $C(a_i)$ is *more informative* than $C(a_j)$ if and only if $\psi^{C(a_j)} > \psi^{C(a_i)}$ for $\psi^{C(a_j)} \leq 0$. Therefore, when comparing the structural density between two clusters, ρ^{random} can be omitted from $\psi^{C(u)}$ and $\psi^{C(u)}$ is simply $\rho^{C(u)}$.

Example 4.2 Suppose we wish to summarize the PPI in Fig. 4.4a into a 3-node summary ($k = 3$). If clusters `apoptosis`, `receptors`, and `TGF-beta` are chosen—instead of the clusters in Fig. 4.4c—we can see that the profit obtained is suboptimal.

Algorithm 1 Algorithm `FUSE`

Input: G, Δ, D, k, b, d, β
Output: Θ_{min}
1: Let S = empty set
2: Let B_{map} = set of pairs (i, b) for each $i \in V$
3: Assign $\psi^{C(u)}$ and $c^{C(u)}$ for each $C(u) \in S_\Delta$
4: $i = 0$
5: **while** $i < k$ **do**
6: $(C_{\text{min}}, B_{\text{map}})$ = **MapProfit**$(S_\Delta, B_{\text{map}}, d, |V|, k)$
7: Remove C_{min} from S_Δ
8: Add C_{min} to S
9: $i = i + 1$
10: **end while**
11: **for** $C(i), C(j) \in S$ **do**
12: **if** $C(i) \neq C(j)$ and $P_i(X > oc_{C(i)C(j)}) \leq 2\beta/|S|^2$ **then**
13: Add edge $(C(i), C(j))$ to F
14: **end if**
15: **end for**

Information budget for proteins `b,c` are depleted due to redundancy, while information budget for proteins `d,e,g,i` are untapped. In contrast, functional summary in Fig. 4.4c is relatively more optimal, as not only the set of clusters maximizes profit through superior coverage and minimal redundancy, but it also maximizes profit through higher structural information (e.g., the cluster `transcription` is structurally dense compared to `apoptosis`).

4.6 The Algorithm FUSE

The profit maximization problem is a variation of the *budgeted maximum coverage problem* [24], which is an NP-hard problem. To permit a tractable solution, let us first consider a straightforward greedy approach. The initial FSG is an empty graph. Given the input PPI network G, $\psi^{C(u)}$ for each functional cluster $C(u) \in S_\Delta$ are computed. The algorithm then iteratively selects the functional cluster that leads to greatest increase in net profit of the summary. Each time a functional cluster $C(u)$ is selected, the FSG and budget information $b(i)$ for every protein $i \in V(u)$ is updated. Once k clusters have been selected, the algorithm terminates by generating the FSG.

A major weakness of the aforementioned approach is that it tends to be "over-enthusiastic" in selection of functional clusters during early iterations. Functional clusters that are too large or too small may be selected at early iterations resulting in very poor cluster choices at later iterations due to limited information budget and summary size (k) constraint. Hence, our proposed algorithm adds a *complexity cost* to each chosen cluster. Given graph size $|V|$ and summary size k, the *expected cardinality* of a functional cluster in the summary is defined by $E[|C|] = \frac{|V|}{k}$. Then the *size deviation cost*, denoted as $c^{C(u)}$, is defined as the square of the deviation of

Algorithm 2 The *MapProfit* procedure.

Input: S_Δ, B_{map}, d, $|V|$, k
Output: C_{min}, B_{map}

1: Let $p_{\text{max}} = 0$
2: **for** $C(u) \in S_\Delta$ **do**
3: Let $B_{\text{temp}} = B_{\text{map}}$
4: Let $p = 0$
5: **for** $i \in V(u)$ **do**
6: Let $(i, b(i)) \in B_{\text{temp}}$ and $p(i) = b(i) - \psi^{C(u)}$
7: **if** $p(i) > 0$ **then**
8: $p = p + \psi^{C(u)}$
9: $b(i) = b(i) - \psi^{C(u)}$
10: **else**
11: $p = p + b(i)$
12: $b(i) = 0$
13: **end if**
14: **end for**
15: $c^{C(u)} = \left(|V(u)| - \frac{|V|}{k}\right)^2$
16: $p = p - c^{C(u)}$
17: **if** $p_{\text{max}} < p$ **then**
18: $p_{\text{max}} = p$
19: $C_{\text{min}} = C(u)$
20: **end if**
21: **end for**
22: **for** $i \in V_{\text{min}}$ **do**
23: Let $(i, b(i)) \in B_{\text{map}}$ and $p(i) = (d/10)(b(i) - \psi^{C(u)})$
24: **if** $p(i) > 0$ **then**
25: $b(i) = (d/10)(b(i) - \psi^{C(u)})$
26: **else**
27: $b(i) = 0$
28: **end if**
29: **end for**
30: **return** (C_{min}, B_{map})

$|C(u)|$ from $E[|C|]$. That is,

$$c^{C(u)} = \left(|V(u)| - \frac{|V|}{k}\right)^2$$

Observe that the greater the difference between $|V(u)|$ and $E[|C|]$, the less likely it is to be part of a summary of k-granularity. Clusters whose size deviate too much from the expected cardinality are penalized and therefore less likely to be selected. This reduces the chance of having too less or too much information budget remaining during the later iterations of the greedy heuristic.

The aforementioned intuition is realized in FUSE as outlined in Algorithm 1. It consists of three phases, namely, the *initialization* phase, the *greedy iteration* phase, and the *summary graph construction* phase. In the initialization phase (Lines 1–3), $\psi^{C(u)}$ and $c^{C(u)}$ for each functional cluster $C(u) \in S_\Delta$ are computed. The greedy iteration phase (Lines 4–10) involves iterative addition of functional clusters into S in a greedy manner as described above. The best candidate functional cluster for the current round ($C_{\min}$) is determined through the subroutine **MapProfit** (Line 6). This step also maintains the information profit of S and the remaining information budget of every v in G through a persistent *profit map* (B_{map}). $C_{\min}$ is then removed from the candidate pool S_{Δ} and added to the solution set S (Lines 7–8). Finally, the summary graph construction phase (Lines 11–15) computes F to generate the FSG $\Theta_{\min}$.

The **MapProfit** procedure is outlined in Algorithm 2. In order to identify the best candidate cluster of the current iteration round, it evaluates every cluster in the candidate pool by evaluating its profit gain potential (Lines 1–21). First, the amount of information to extract from each protein's information budget pool ($b(i)$) is computed (Lines 7–13). Next, the potential profit gain is adjusted to compensate for the complexity cost (Lines 15–16). After $C_{\min}$ is found, the profit map is recomputed to commit the changes made to the information budget map due to the selection of $C_{\min}$ (Lines 21–29).

Theorem 4.1 *Algorithm* FUSE *takes* $O(|S_{\Delta}|^2|V|^2)$ *time in the worst case.*

Proof In the initialization phase, $\psi^{C(u)}$ for each $C(u) \in S_{\Delta}$ has to be computed. Each $C(u)$ may contain up to $|E|$ edges and $|V|$ vertices. In Algorithm 1, $\psi^{C(u)}$ for each $C(u) \in S_{\Delta}$ takes $O(|E|)$ time. Thus, the total complexity for this procedure is $O(|E||S_{\Delta}| + |V||S_{\Delta}|)$ time.

In the greedy iteration phase, the **MapProfit** subroutine defined in Algorithm 2 is evaluated k times. In Algorithm 2, Lines 2–21 require $O(|S_{\Delta}||V|)$. Lines 22–29 require $O(|V|)$ time. Thus, Algorithm 2 takes $O(|S_{\Delta}||V| + |V|)$ time. The iteration phase, as such, takes $O(k|S_{\Delta}||V| + k|V|)$ time in total.

Finally, the summary graph construction phase involves pairwise significance evaluation of the resultant functional cluster set. This involves evaluation of all edges between k-pairwise functional clusters of the summary. Each significance $P_i(X > oc_{uv})$ test requires a single-pass evaluation of edges connecting a pair of clusters. At worst case, this takes $O(|E|)$ time. The summary graph construction phase therefore require $O(k^2|E|)$ time.

The FUSE algorithm, as whole, takes $O(|E||S_{\Delta}| + |V||S_{\Delta}| + k|S_{\Delta}||V| + k|V| + k^2|E|)$ time. In the worst case scenario of $|E| = |V|^2$ and $k = |V|$, the algorithm takes $O(|S_{\Delta}||V| + |S_{\Delta}||V|^2 + |V|^2 + |V|^4)$ time, implying a polynomial time complexity at worst case.

Example 4.3 Consider as an example the summarization of the PPI in Fig. 4.4a. Lines 1–3 in Algorithm 1 construct the candidates shown in Fig. 4.4b and compute, for each candidate $C(u)$, its structural information value $\psi^{C(u)}$ and cost $c^{C(u)}$. Following that, the modified greedy iteration phase selects k candidates by profit maximization (Lines 4–10 in Algorithm 1). Figures 4.4c, d show examples of functional subgraphs selected

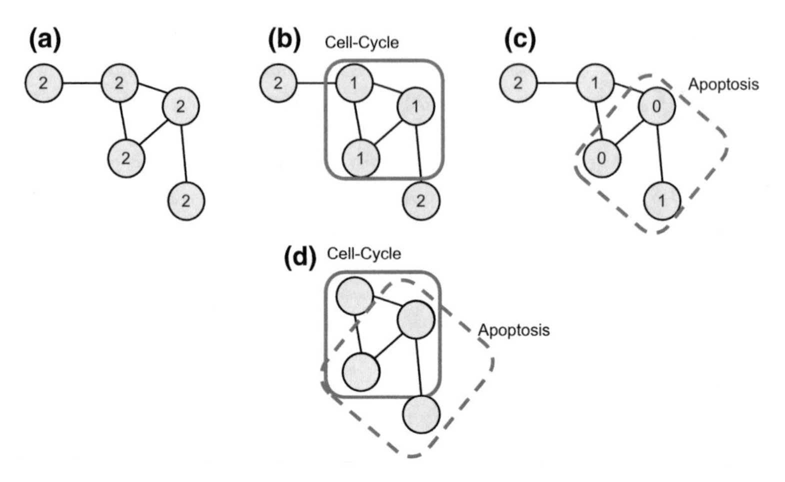

Fig. 4.6 Illustration of **MapProfit** procedure

following the greedy iteration phase. Finally, the edges depicted in Fig. 4.4c, d that indicate the functional relationship between the functional subgraphs are computed in Lines 11–15 in Algorithm 1.

Figure 4.6 further illustrates the **MapProfit** procedure in Algorithm 2. Figure 4.6a shows a toy PPI network with each protein assigned an initial information budget of $b = 2$. Figure 4.6b shows the selection of the `cell-cycle` functional subgraph with structural information value $\psi^{C(u)} = 1$. Observe that the information budget remaining for each affected protein is updated accordingly. Figure 4.6c shows the remaining information budget when another functional subgraph (`apoptosis` with $\psi^{C(u)} = 1$) is selected. Finally, Fig. 4.6d depicts the summary of functional subgraphs selected.

4.7 Experimental Results

FUSE is implemented in Scala and Java. We now present the experiments conducted to evaluate the performance of FUSE and report the results obtained. The PPI datasets employed in this study are shown in Table 4.2. *Biological Process* (BP), *Molecular Function* (MF), and *Cellular Component* (CC) GO annotations are used. Unless specified otherwise, we set $\beta = 0.01$, $b = 3$, and $d = 0$ in order to balance *coverage* and *redundancy* of the functional summaries. We assign all edge weights be 1.0. All experiments were run on a 1.66 GHz Intel Core 2 Duo T5450 machine, with 3 GB memory, and a 250 GB SATA disk.

Table 4.2 Summary of datasets used

Dataset	#nodes	#edges	Source
H. sapiens	9181	34624	HPRD [25]
S. cerevisiae	4768	177299	IntAct [3]
D. melanogaster	3114	6472	IntAct
Alzheimer's disease (AD)	177	1038	IntAct

4.7.1 Evaluation Metrics

Coverage. We use the *coverage* metric to evaluate the fraction of the annotated PPI network covered by a summary. A functional summary with high coverage is desirable because it is more representative of the underlying interaction network than a summary with low coverage. The coverage of a functional summary Θ is defined as:

$$coverage(\Theta) = \frac{\left|\bigcup_{C(u)\in S_\Theta} V(u)\right|}{\left|\bigcup_{C(u)\in S_\Delta} V(u)\right|} \tag{4.5}$$

The coverage is the ratio of the total number annotated proteins in the summary over the total number of annotated proteins in the protein interaction network.

Redundancy. The *redundancy* metric is the average number of functional clusters each protein belongs to. This is an indicator of the amount of cluster overlap in the summary. Redundancy of Θ is defined as:

$$redundancy(\Theta) = \frac{\sum\limits_{C(u)\in S_\Theta} |V(u)|}{\left|\bigcup\limits_{C(u)\in S_\Theta} V(u)\right|} \tag{4.6}$$

A summary Θ with no overlapping clusters will have lowest possible redundancy value of 1, where every protein is assigned to exactly one cluster. A summary with high redundancy is undesirable, because a summary with many highly overlapping clusters is less intuitive and more complicated.

Precision and Recall. The following well-known evaluation metrics are also used – *precision and recall*. These are well known statistical measures to indicate accuracy and completeness. As discussed earlier, precision, a measure of exactness, is defined as:

$$precision = \frac{true\ positive}{true\ positive + false\ positive} \tag{4.7}$$

Recall, a measure of completeness, is defined as:

$$recall = \frac{true\ positive}{true\ positive + false\ negative} \quad (4.8)$$

If a cluster $C(i)$ is assigned with the function i, then any protein $p \in C(i)$ that is not annotated with i or its descendants is deemed a false positive. If $p \in C(i)$ is annotated with i or descendants, it is a true positive. Likewise, a protein $p \in V$ that is annotated with i but not in $C(i)$ is deemed a false negative. Here, proteins without annotations are not taken into consideration.

4.7.2 **FUSE** *Versus Graph Clustering Methods*

Dataset. Currently, there does not exist any gold standard to compare functional summaries of PPIs. Typically, biological graph clustering approaches use MIPS complex annotations [26] as gold standard data for testing cluster quality. These annotations, however, are limited to complexes and not for other functional clusters like pathways. GO annotation data is also used as gold standard for evaluating clustering algorithms. As our approach utilizes attributes of GO, using GO annotations as gold standard evaluation may lead to results biased in favor of FUSE. Instead, we obtained a *different* set of curated attributes as gold standard–the *molecule class* annotations from HPRD–which is distinct from GO attributes. Note that these annotations are only available in the *H. sapiens* dataset. Consequently, we use this dataset for the comparative study. To create a gold standard *reference summary*, we generated a network from subgraphs induced from the HPRD network using nodes grouped by their *molecule class* attribute, signifying the molecular functional groups within the network. Subgraphs from five functional groups corresponding to subgraphs of proteins classified as `G protein coupled receptor`, `Protease inhibitor`, `RNA binding protein`, `Cytoskeletal associated protein`, and `Calcium binding protein` are extracted and merged to form the reference summary network (747 nodes, 959 edges). FUSE and state-of-the-art graph clustering methods are then evaluated on this network to determine whether the graph can be partitioned and summarized to reconstruct the gold standard functional groups.

We compare the performance of FUSE with four popular graph clustering methods for life sciences applications, namely Markov clustering (MCL) [27], MCODE [5], and NeMo [7]. We also compare FUSE with CSV [10], a cohesive subgraph visualization method. Note that in order to obtain higher order modules of a PPI network, the current approach is to first use an existing graph clustering method on the network to generate the clusters followed by function assignment. For example, in Krogan et al. [27], the global yeast PPI network is first clustered using MCL to generate non-overlapping clusters. Then, each cluster is compared against MIPS complex annotations [26] and the complex annotation with the greatest overlap is assigned to represent the cluster.

Cluster quality comparison. We first emphasize on the qualities of an ideal summarization. First, the generated clusters have to be representative of the underlying graph, which implies that coverage of the clustering should be sufficiently high. Second, *attribute purity* [28] of the clusterings should correspond to the functional groups that were merged *apriori*. This can be determined through the purity of the `molecule class` attribute within the proteins in each cluster. Each functional group should also be well-represented. We use *precision*, *recall*, and *F-measure* to quantify these features. For each cluster, we determine the `molecule class` functional group that best matches the cluster. The *purity* of that cluster is then defined as the proportion of nodes in the cluster that belong to the best matching group. As a functional group may be represented by several smaller clusters, we define *recall* for each functional group as total coverage of the functional group among the clusters that best matches that functional group. Then, the *precision* of a clustering is defined as the average purity among all clusters. The *recall* of a clustering is defined as the average recall among all functional groups. Lastly, the *F-measure* ($\frac{2*precision*recall}{precision+recall}$) provides an overall measure of clustering quality.

Figures 4.7, 4.8, 4.9 and 4.10 depict the results of summarization quality by F-measure, precision and recall. Where applicable, we adjust relevant parameters to vary the cluster granularity. As NeMo has no parameter to tweak, only a single set of clusters can be obtained. In MCL, CSV, and MCODE, the *inflation*, η_{mseen} *cutoff*, and *node score cutoff* parameters are adjusted, respectively, to vary the cluster sizes (denoted as k in all figures). In FUSE, the parameter k directly affect the summary granularity. Here, we use k to represent the number of clusters obtained by each method. Because most methods indirectly affect this via parameters, it may not be possible to cover the entire range of possible k values.

Observe that FUSE generates summary with significantly higher F-measure score compared to the graph clustering-based approaches for all values of k. In other words, FUSE may generate summaries at multiple levels of complexity while remaining representative of the underlying graph. Observe that, although NeMo, CSV, and MCODE generate clusters with high precision, the recall scores are very low (< 0.2). This

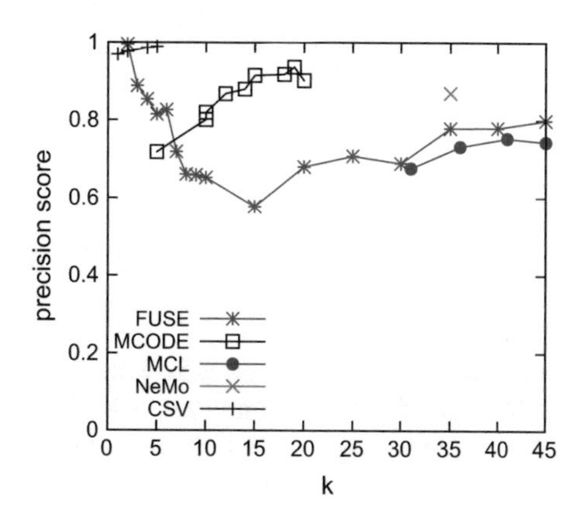

Fig. 4.7 Cluster quality of FUSE versus graph clustering-based approaches (precision)

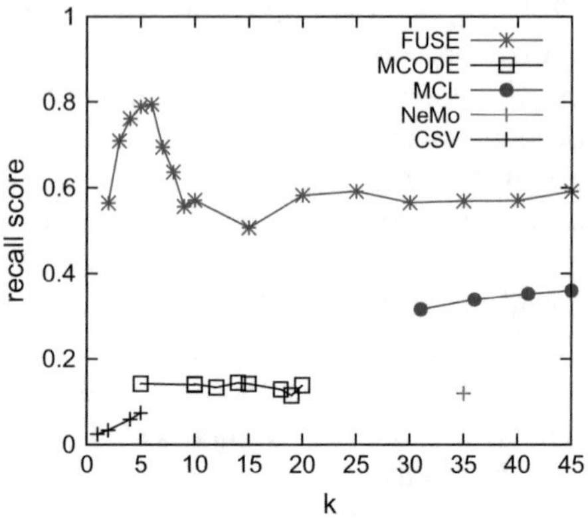

Fig. 4.8 Cluster quality of FUSE versus graph clustering-based approaches (recall)

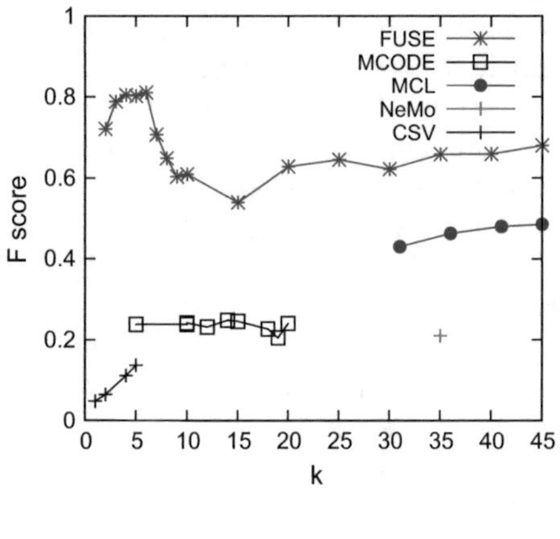

Fig. 4.9 Cluster quality of FUSE versus graph clustering-based approaches (F-score)

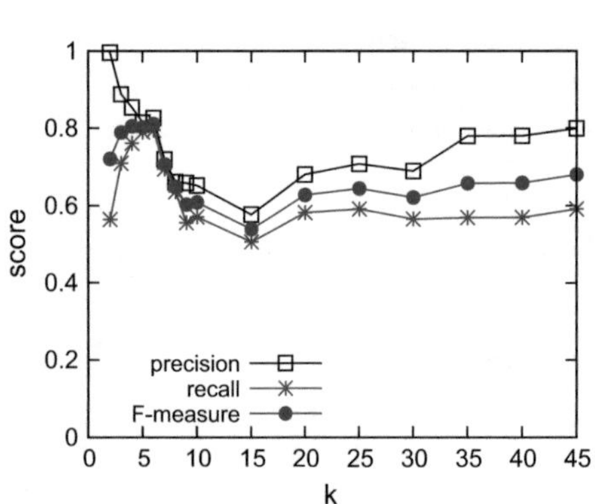

Fig. 4.10 Cluster quality of FUSE versus graph clustering-based approaches (FUSE)

is because these two approaches identify highly cohesive subgraphs, which tend to be part of protein complexes. CSV in particular are limited to identification of near-clique structures. Proteins in complexes belong to the same functional groups and hence the high precision. However as mentioned earlier, biological networks are not comprised solely of complexes. Consequently, majority of the underlying network is poorly represented by these approaches due to heavy bias towards complexes. Specifically, most of the clusters match the RNA binding protein class of proteins, leaving other groups barely represented. For instance, the Protease inhibitor subgraph is not well represented because of its inherent loose structure. Although the recall score of MCL is relatively higher as this method is known to perform very well in biological clustering applications, it is still below 0.4. Note that the MCL approach failed to partition the underlying network into five clusters representing the five functional groups. The CSV approach, on the other hand, were not able to generate larger number of partitions.

Notice that these existing approaches indirectly affect the summary complexity whereas FUSE allows direct adjustment of summary size, which explains why summaries at any level of detail can be obtained by the latter. Figure 4.10 shows that FUSE generates summaries at different granularity without greatly affecting the precision and recall of the clusterings. The peak F-measure score of 0.8 is obtained in FUSE at $k = 5$, corresponding to the five gold standard functional groups that comprise the dataset. Observe that the recall and precision scores are equally high. As cluster granularity deviates from the underlying five functional groups, obviously the F-measure score drops.

Function representativeness comparison. The accuracy and representativeness of the function assigned to each cluster is key to generating high quality functional maps. Here, we introduce measures that quantify the representativeness of functions assigned to each clusters and compare FUSE to graph clustering methods in this aspect.

To obtain the functional landscape of a PPI, graph clustering methods often assign function to clusters through functional enrichment techniques. To this end, we compute the statistical significance of association of the cluster with every GO term based on the hypergeometric distribution [4]. The term with the best *p-value* is assigned as the *representative function* of the cluster. To evaluate the representativeness of this assigned function, we reuse the precision and recall measures introduced earlier with slight modification. Specifically, the *representative purity* of a cluster is defined as the proportion of nodes in the cluster that are annotated with the representative function. We also define *representative recall* for each functional group as total coverage of the functional group among the clusters that has the functional group assigned as representative function. Then, the *precision* of the representative functions is defined as the average representative purity among all clusters, and the *recall* of the representative functions is defined as the average representative recall among all functional groups.

Figures 4.11 and 4.12 depict the representativeness of the functional summaries by different techniques. As FUSE is designed specifically to generate highly representative maps, each cluster is perfectly representative of the biological function assigned to it. Likewise, each function is well represented by its assigned cluster. In graph clustering methods, however, the clusters do not represent their representative function well, as indicated by the lower precision score. Hence, proteins within the

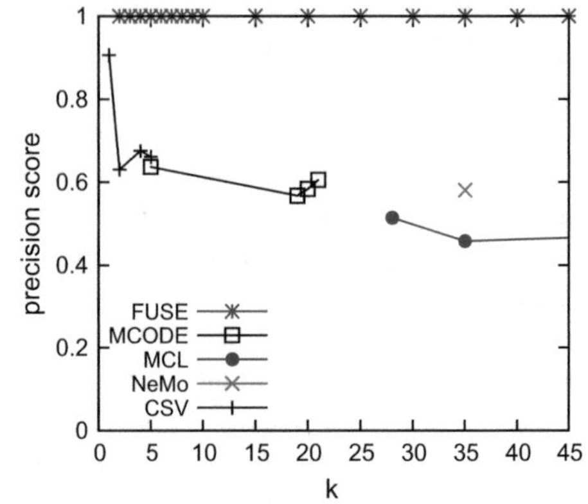

Fig. 4.11 Function representativeness (precision)

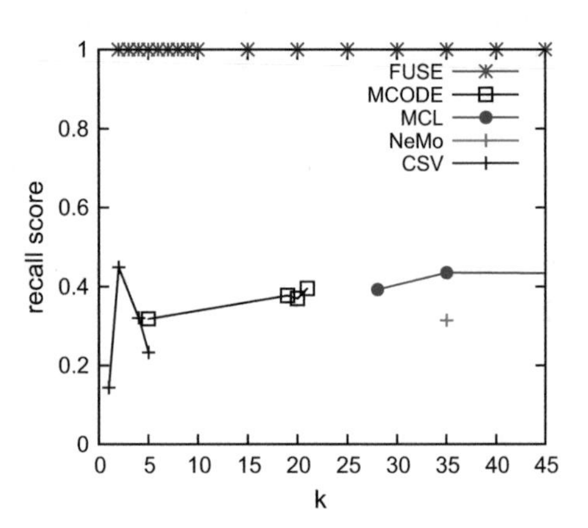

Fig. 4.12 Function representativeness (recall)

clusters exhibit less functional coherence. The lower recall scores in graph clustering methods imply that only a fraction of nodes annotated with the representative function are included in the cluster. That is, FUSE summaries contain functional clusters that are more representative of the assigned function, and thus provide more meaningful and interpretable higher-order functional maps of the underlying PPI. While clusters without attribute coherence may still reveal novel biological insights, assigning a function to represent such cluster could be misleading.

Qualitative evaluation. Next, we qualitatively compare the summaries generated by both approaches for the DNA *S.cerevisiae* dataset. We argue that functional summaries are best evaluated qualitatively, partly because of the lack of a gold standard dataset for higher order function-function associations. We chose small and functionally specific subnetwork rather than large global networks so that qualitative comparison is feasible. To this end, we extracted the subnetwork containing DNA replication related proteins of *S.cerevisiae* network in `IntAct` ($n = 105$) as evaluation dataset. This dataset is obtained from the *S.cerevisiae* global network by extracting the induced subgraph whose proteins share `DNA-dependent DNA`

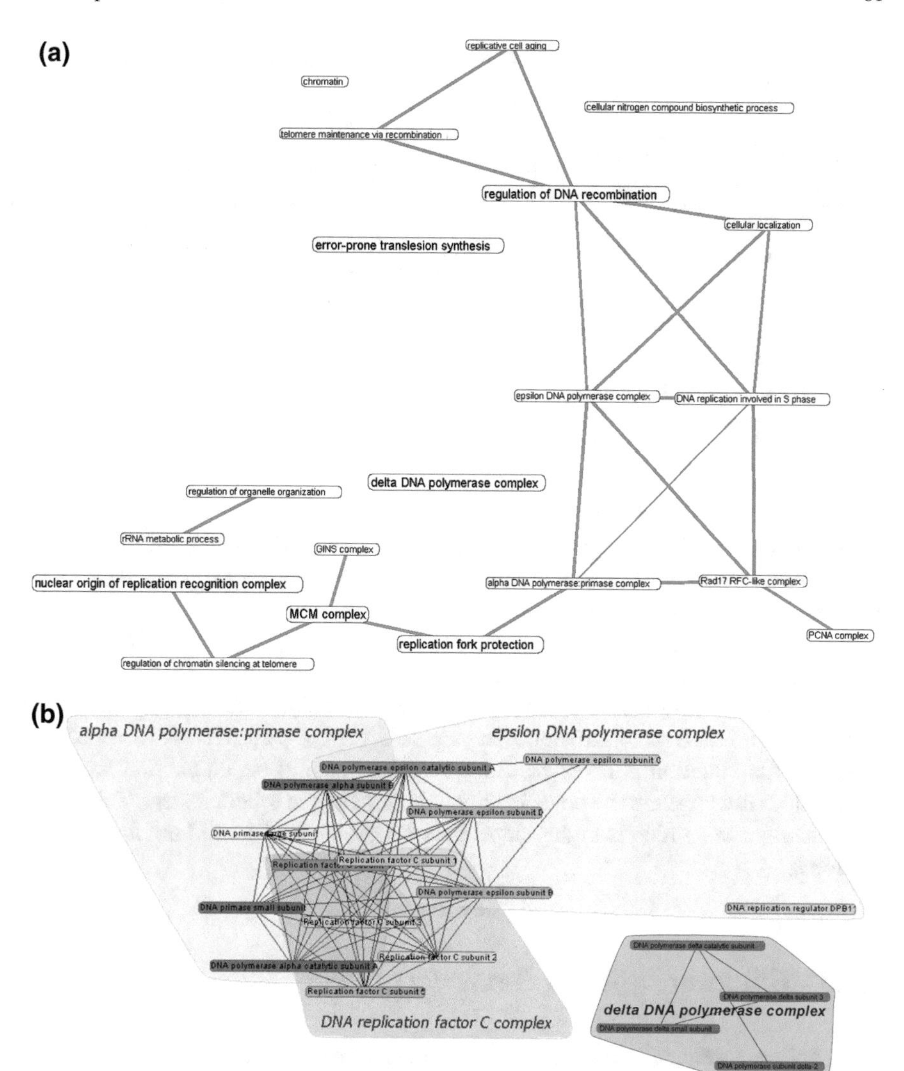

Fig. 4.13 Functional summarization of DNA *S.cerevisiae*. **a** Summary at $k = 20$ obtained through FUSE. Nodes represent functional clusters. The size of a node correlates with the number of proteins that constitute the functional cluster. Edges represent associations between functional clusters; the stronger the association, the thicker the lines. **b** Underlying protein interaction networks of α-DNA polymerase, ε-DNA polymerase, δ-DNA polymerase, and DNA replication factor C complex

replication function. Here, we compare our results qualitatively against the MCL approach. Figure 4.13a shows the FUSE generated functional map and Table 4.3 shows the MCL generated clusters. FUSE is able to partition the network into major components of DNA replication. Critically, DNA polymerase complexes ($\alpha, \delta, \varepsilon$) – key components in DNA replication – are obtained. MCL, on the other hand, is not

Table 4.3 Summary of DNA *S.cerevisiae* obtained through Cluster+Enrich (9 single member clusters are excluded)

Cluster	Precision	Recall
Replication fork protection complex	0.8	0.67
Postreplication repair	0.8	0.8
DNA recombination	0.77	0.53
Nuclear origin of replication recognition complex	0.86	1.0
GINS complex	0.43	0.75
Negative regulation of cell cycle process	0.6	0.75
Alcohol metabolic process	0.5	0.5
DNA replication factor C complex	0.22	1.0
rRNA metabolic process	1.0	0.5

able to obtain the polymerases. Deeper analysis reveals that many proteins in the *DNA replication factor C complex* cluster of Cluster+Enrich actually belong to DNA polymerases. This is a misrepresentation.

As shown in Fig. 4.13b, proteins in `DNA polymerase complexes` α and ε and `DNA replication factor C complex` strongly interact with each other, forming a tight clique (with exception of 2 proteins). Hence, they cannot be separated via graph structure alone. In case of δ `DNA polymerase complex`, however, the situation is reversed. As it contains incomplete interaction data, the cluster does not appear to be densely interacting relative to other clusters. This could explain why MCL, which is highly dependent on structural data, did not identify the complex.

4.7.3 *Effects of Different Parameters*

Effect of parameter k. Recall that the user-defined parameter k controls the granularity of the summary. Intuitively, as k increases the amount of information contained within the summary as well as its complexity increase. Fig. 4.14 reports the effect of k on the summaries of test datasets. As k increases, the *summary information content* (SIC), denoted by $SIC(\Theta)$, rises rapidly until it saturates to a peak value before tapering off.

$$SIC(\Theta) = \sum_{C(u) \in S_\Theta} -\psi^{C(u)} |V(u)| log p_{V(u)} \qquad (4.9)$$

where $p_{V(u)}$ is the probability that a protein in network is annotated with term u or its descendants. Note that summary profit cannot be used for comparing summaries with different k values because it does not make any assumption about the information

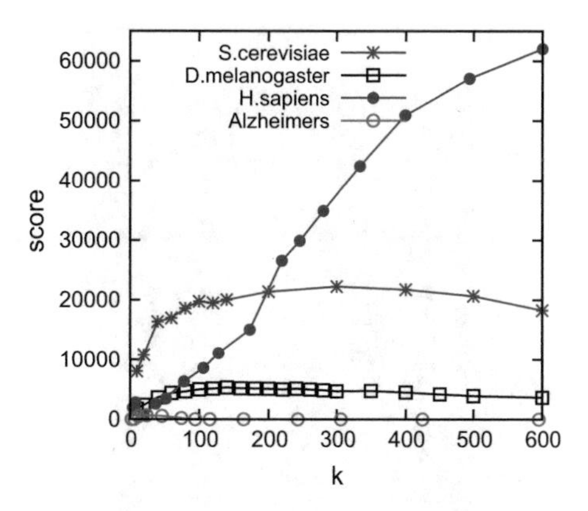

Fig. 4.14 Effect of k on summary SIC

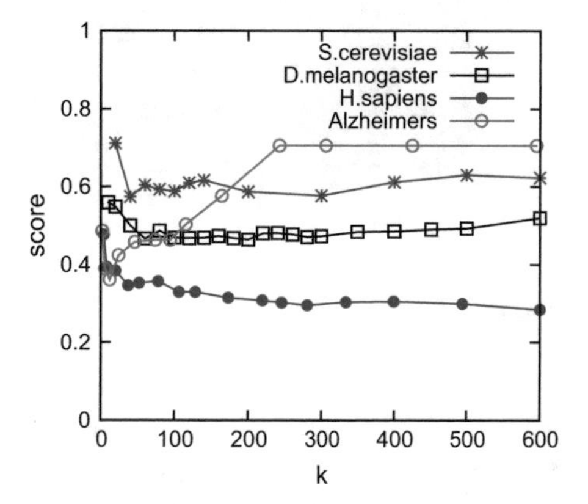

Fig. 4.15 Effect of k on summary coverage

content of a GO term attribute. In contrast, SIC measure is summary profit with a twist – small clusters are weighted higher than large clusters. This allows one to compare information content of summaries with different k values. Other factors being equal, a summary with many small clusters will contain more information than a single large cluster. The above results imply that k is useful up to a certain value, after which increasing k only increases summary complexity while providing little information gain.

Figure 4.15 plots the effect of k on coverage of the summary. Observe that except for low k values, it is relatively stable as k varies. In fact, the amount of information a summary can provide is limited by the resolution and completeness of the interaction and annotation data. This could explain why *S. cerevisiae* summaries have consistently higher coverage and information content than *D. melanogaster* summaries. The *H. sapiens* summary contains the largest number of nodes and edges, and even at $k = 600$, information content is still increasing. The smaller AD network, however, reaches a peak of information content at $k = 20$.

Effect of parameters b and d. We investigate the effect of user-defined parameters b and d on summary coverage and redundancy. We use the global *S. cerevisiae* dataset

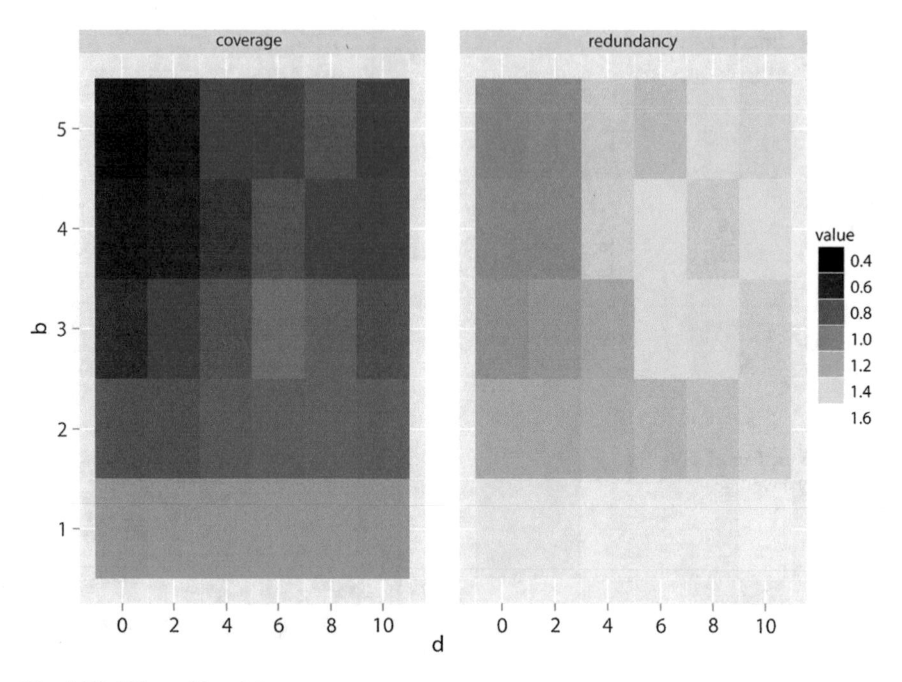

Fig. 4.16 Effect of b and d

with $k = 100$. Figure 4.16 shows that increasing b or decreasing d lowers overall summary redundancy at the expense of lower summary coverage. On the other hand, when d is increased or b is decreased, both summary redundancy and coverage increases. An intuitive explanation of this phenomenon is that more cluster overlap penalty means fewer combination of clusters to choose from, and therefore lower likelihood of finding a combination of clusters with high coverage. Both parameters allow users to control the coverage and redundancy trade-off, depending on whether it is preferable to have more coverage or less redundancy (Fig. 4.17).

Statistical significance. We now evaluate the statistical significance of a FUSE generated summary. Evaluation of graph clusters is not trivial because there is no analytic solution for the exact p-value of a cluster. However, an upper bound can be computed to detect if the density of a subgraph of a given size is statistically distinct compared to one that is randomly constructed. Given the graph G, we utilize the following upper bound derived from [29] as the $p - value$ of a functional subgraph cluster:

$$P(R_{\rho_1} \geq (1+\varepsilon)\log n/\kappa(\rho_1, \rho)) \leq \frac{(1-\rho)^{0.5}}{2\pi\rho^{0.5}} \frac{(1+\varepsilon)\log n}{n^{\varepsilon(1+\varepsilon)\log n/\kappa(\rho_1,\rho)}} \tag{4.10}$$

where:

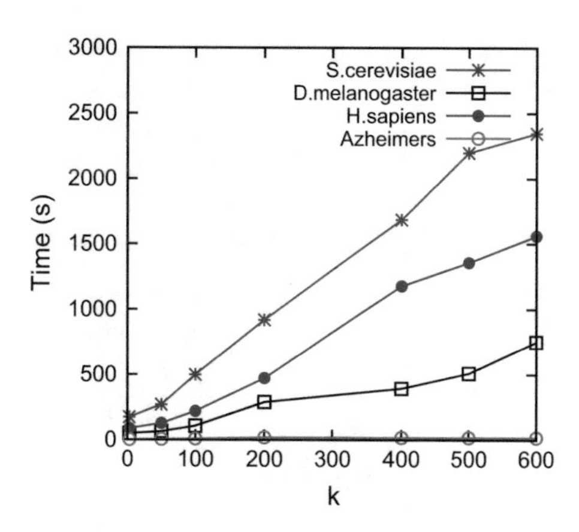

Fig. 4.17 Running times of FUSE (in sec.)

$$\varepsilon = \frac{r - \log n/\kappa(\rho_1, \rho)}{\log n/\kappa(\rho_1, \rho)} \tag{4.11}$$

$$\kappa(\rho_1, \rho) = \rho \log \frac{\rho}{\rho_1} + (1-\rho) \log \frac{1-\rho}{1-\rho_1} \tag{4.12}$$

where ρ is the expected probability of observing an edge between two nodes, R_ρ is the size of the maximum subset of vertices that induce a ρ-dense subgraph, r is the subgraph size, and n is $|V|$.

Using the $p - value$ above, we compute the $p - value$ upper bound of a given FUSE cluster and the associated cluster size needed to satisfy the upper bound. In Table 4.4, we show the upper bounds of $p - value$ significance of at most 0.05 and the cluster size needed to satisfy the bound. Observe that all of the clusters we obtain from FUSE summary are at least as large as the required size. Thus, these clusters have $p - values$ that are significant. One weakness of the above formulation is that not all clusters can be associated with a meaningful $p - value$ upper bound (there are bounds larger than 0.05, thus they cannot be used to meaningfully assess significance).

Effect of annotation loss. Next, we evaluate the effect of loss of annotations on FUSE algorithm. We observe how a summary changes when annotations are gradually removed from the PPI network. To achieve this, we first let Θ_0 be the FUSE summary generated using the full annotation dataset. Next, we remove a fraction of the annotations. Let the *annotation loss rate* be the fraction of annotations removed. For example, the annotation loss rate of 0.5 implies that half of the annotations in the PPI network has been removed. Given this measure, we compared FUSE summaries of annotation loss rate from 0.05 to 1.0 against Θ_0 generated from the human PPI network.

To measure the similarity of a pair of summaries, we employ the *Jaccard index* (JI) [30] evaluation measure. Given two summaries Θ_i and Θ_j, the *Jaccard index* is defined as $J(\Theta_0, \Theta_j) = \frac{A}{A+B+C}$, where A is the number of protein pairs that is

Table 4.4 The p-value significance of FUSE clusters

Cluster size	Cluster size to satisfy p-value	p-value
4	1.940572357	1.99E-06
5	2.381391052	1.83E-05
4	2.51869916	3.31E-05
4	2.51869916	3.31E-05
3	2.51869916	3.31E-05
3	2.51869916	3.31E-05
3	2.51869916	3.31E-05
3	2.51869916	3.31E-05
3	2.51869916	3.31E-05
3	2.51869916	3.31E-05
7	2.740737755	7.99E-05
9	2.781159841	9.31E-05
5	2.839581494	1.16E-04
5	2.839581494	1.16E-04
5	2.839581494	1.16E-04
5	3.485484996	9.72E-04
5	3.485484996	9.72E-04
14	3.810967312	0.002456578
10	3.97437195	0.003801536
6	4.467241236	0.012844867
8	4.557278228	0.015816221

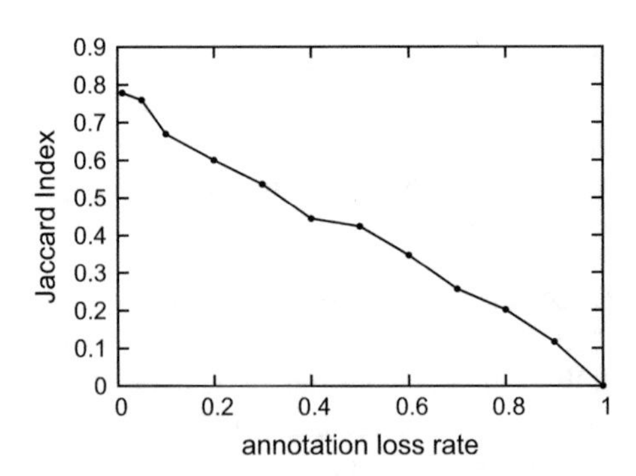

Fig. 4.18 Effect of annotation loss

co-clustered in both Θ_i and Θ_j, B is the number of protein pairs co-clustered in Θ_i but not Θ_j, and C is the number of protein pairs co-clustered in Θ_j but not Θ_i. $J(\Theta_i, \Theta_j) \in [0, 1]$ and $J(\Theta_i, \Theta_j = 1$ if the summaries are identical.

Figure 4.18 shows the effect of annotation loss on FUSE summaries. We observe that the Jaccard index similarity between Θ_0 and a summary with annotation loss rate gradually decreases as more annotations are removed. The gradual drop in similarity suggests that our approach is robust against loss of annotation.

4.7.4 Runtime and Scalability

Figure 4.17 plots the running times of FUSE over the real datasets for generation of summaries ranging from $k = 3$ to $k = 600$. Observe that it increases almost linearly with k. Specifically, summarization of the yeast interaction network (the largest available network) completes within 40 min for $k = 600$. For practical sizes of $k = 3$ to $k = 100$, a functional summary of a PPI can be generated within few minutes. Disease networks such as AD network can be completed in less than 10 s.

We now assess the scalability of FUSE with respect to network size and $|S_\Delta|$. Note that the latter feature is important as it will continue to grow as more annotation information becomes available. To assess the scalability with respect to network size, we generated synthetic networks of vertex size $|V| = 100$ to $|V| = 20000$. Note that the largest available PPI (human network) has only around 9000 vertices. For every term t, a vertex has a 2% probability of being annotated with it. The number of terms is $|S_\Delta| = 2769$. The *edge density* of the synthetic networks is such that the probability that a pair of vertices interact is 0.0025, resulting in an average of 1 million edges in a network of 20000 vertices. Summary granularity is set to $k = 50$. To measure the effect of $|S_\Delta|$ on running time, we generated synthetic networks by varying $|S_\Delta|$ ranging from $|\Delta| = 100$ to $|\Delta| = 10000$.

Figures 4.19 and 4.20 depict the scalability of FUSE with respect to $|V|$ and $|S_\Delta|$. As the number of vertices increases, the execution time of FUSE increases in a quadratic fashion. In fact, it appears to increase almost linearly for networks with $|V| < 10000$. For larger networks, the $\psi^{C(u)}$ component and the FSG generation component take up the bulk of the execution time. Observe that in Fig. 4.20, the FSG generation component takes up bulk of the computation time and is independent of $|S_\Delta|$. As $|S_\Delta|$ increases, $\psi^{C(u)}$ computation and iterative cluster selection time increases in near linear fashion, demonstrating ability of FUSE to handle high-dimensional annotation data.

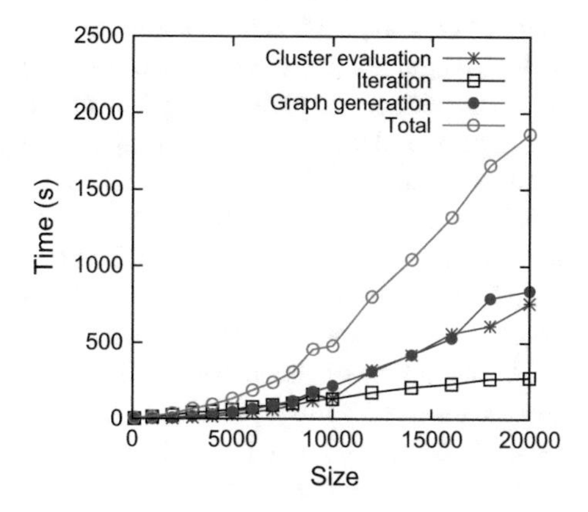

Fig. 4.19 Scalability of FUSE versus annotation size

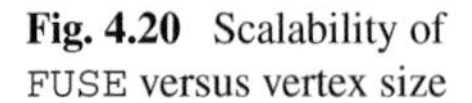

Fig. 4.20 Scalability of FUSE versus vertex size

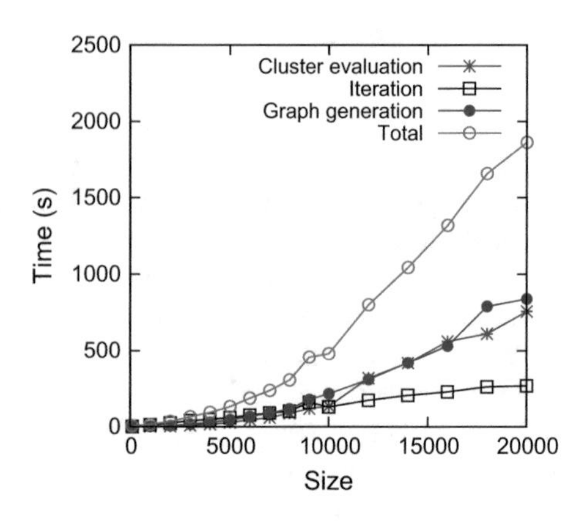

4.8 Case Study on AD Network

In this section, we construct a low and a high resolution functional summaries of the AD network to illustrate the benefits of FUSE in providing a higher level functional view of the underlying PPI. A low resolution summary delineates broad functional overview of the processes related to the disease whereas a high resolution summary provides in-depth functional landscape of the disease, revealing associations between processes related to the disease. Figure 4.2 shows a low resolution summary ($k = 10$) of the AD network. It indicates that the AD network is represented by an interconnection of several key processes, include protein phosphorylation *(B7)*, cell-cell signaling *(B2, B3)*, and microtubule-based transport and localization *(B1, B5)* processes.

Figure 4.1 depicts a high resolution functional summary for $k = 30$. Defective transport mechanism has major implications in AD. Consequently, several transport and cytoskeleton organization related cellular processes are represented in the summary *(A11, A22, A24, A26)*. Disrupted transport mechanism affects, among others, synapse organization and vesicle trafficking *(A6, A8, A23)*. In the literature, several lines of evidence explain disruption of transport and its related processes in AD. `Amyloid-`β plaques may lead to hyperphosphorylation of `tau proteins`, subsequently causing microtubule defects and axonal transport impairment [31]. More strikingly, recent findings indicate that vesicle transport itself play a causative role in pathogenesis of the disease [32]. `Glucose metabolic processes` *(A20)* is closely linked to microtubule-based processes *(A22, A24)*. The link between bioenergetics and transport in AD has been discussed in [33].

At the center of the summary lies `protein folding` and `calcium ion homeostasis` pathways *(A15,A17)*. Protein misfolding is central to AD pathogenesis [34]. Misfolded `amyloid-`β accumulation is shown to induce calcium overload, leading to a variety of structural and functional disruption in neurons [35]. The two functional clusters are among the nodes with the highest degree in the summary. Cell fate processes that trigger or inhibit differentiation and cell fate *(A9, A10, A12)* are also linked to AD [36]. It has been suggested that `Wnt signaling` dysregulation, a key developmental pathway, leads to reduced synaptic plasticity and function in AD [37]. Processes such as peptide cross-linking and negative regulation of angiogenesis *(A3, A4)* imply vascular roles in AD pathogenesis [38].

From signaling regulation perspective, five major signaling pathways are implicated – small `GTPase` *(A28)*, `Notch` *(A14)*, `Wnt receptor` *(A18)*, `glutamate` *(A21)*, and `G-protein coupled receptor` signaling pathways *(A16)*. Several functional clusters connect with multiple signaling pathways, indicating that signaling pathways crosstalk in AD pathogenesis. For instance, the `serine/threonine kinase GSK-3`β, a potential therapeutic target, is known to be regulator of both the `G-protein coupled receptor` pathway and the `Wnt/`β`-catenin` signaling pathway [39]. `PS1` may be involved in regulating both `Notch` and `Wnt` pathways in AD [40].

The tight interplay of multiple pathways and processes in the aforementioned functional summary of AD network highlights the complexity of the disease. The disease remains poorly understood despite decades of research. While the summary does not suggest causal relationships, in part because of the undirected nature of the FSG, we hope that by having a global, big picture view of process-process interactions, researchers can better identify the causative mechanisms of AD. Most studies considered an aspect of the processes in isolation. An integrative study, however, may lead to a more consistent view of the disease that addresses distinct, often competing hypotheses (Table 4.5).

Table 4.5 High-degree CC functional clusters in the *H. sapiens* summary ($k = 400$)

CC Functional Cluster	Degree
Heterogeneous nuclear ribonucleoprotein complex	183
Cytosolic large ribosomal subunit	161
Cytosolic small ribosomal subunit	158
Coated pit	158
Mitochondrial nucleoid	149
Chaperonin-containing T-complex	148
CRD-mediated mRNA stability complex	141
NuA4 histone acetyltransferase complex	136
Actin filament	135
Actomyosin	134
Clathrin coat of coated pit	133
Nonhomologous end joining complex	124
Endocytic vesicle membrane	124
Nucleosome	124
Nuclear inner membrane	123

4.9 Inferring Functional Cluster Hubs

Structural information provided by the summaries presents an opportunity to study the topology and connectivity of higher order abstractions of the underlying PPI network. Here we analyze the association patterns of functional clusters in summaries of the global *H. sapiens* PPI. To this end, we generate cellular component (CC) and biological process (BP) summaries of the human network. For each summary type, we vary the level of detail by setting k from 50 to 400.

Figure 4.21 shows the frequency-degree plots of the functional clusters at different k values. At the broadest level of abstraction ($k = 50$), long-tailed degree distribution of functional clusters is not observed. As level of detail increases to $k = 400$, the smaller and more specific clusters exhibit increasingly pronounced long-tailed distribution characteristics. We note that the CDF plots on a semi-log scale form straight lines at higher k values ($k = 200$ and $k = 400$), implying exponential distribution of the cluster degrees.

In light of heavy-tailed distribution of functional cluster degrees at higher k values, we identify *functional cluster hubs* in the summary of the human network ($k = 400$) (analogous to identification of protein hubs). While Patil and Nakamura defined hub as proteins having degree of more than 6 [41], we chose a higher threshold such that they correspond to the 15 most connected functional clusters. The list of functional hubs is shown in Table 4.6.

We observe that CC cluster hubs in *S. cerevisiae* can be categorized into several major functional groups. A significant percentage of the cluster hubs – such as `cytosolic large ribosomal subunit`, `cytosolic small`

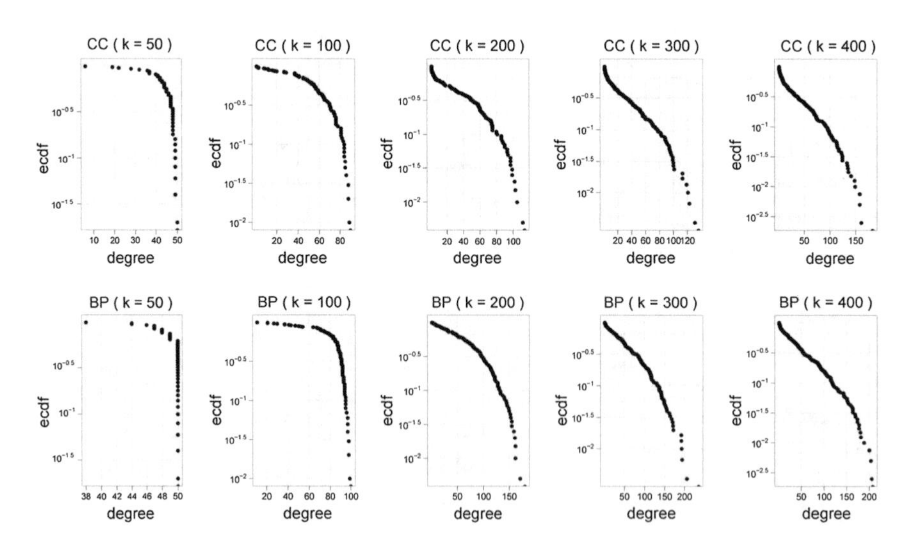

Fig. 4.21 Connectivity of functional clusters in *H. sapiens* network. Functional cluster degree CDF plots for BP and CC summaries at varying cluster granularity. Plots are on a semi-log scale

Table 4.6 High-degree BP functional clusters in the *H. sapiens* summary ($k = 400$)

BP Functional Cluster	Degree
Actin filament bundle assembly	208
Regulation of defense response to virus by virus	206
Negative regulation of catabolic process	204
Peptidyl-threonine phosphorylation	200
Signal complex assembly	189
Positive regulation of protein complex assembly	182
Regulation of nitric oxide biosynthetic process	181
Glial cell development	178
Cell killing	178
Regulation of cytokine-mediated signaling pathway	174
Protein stabilization	174
Actin filament capping	170
Activation of MAPKK activity	169
T cell receptor signaling pathway	164
Regulation of RNA splicing	164

`ribosomal subunit`, `eukaryotic translation initiation factor 4F complex`, `preribosome, small subunit precursor`, `preribosome, large subunit precursor`, and `polysome`– are core to regulation and functioning of protein translation. It is unsurprising that these functional clusters have high degree, since every protein must be translated or regulated by these machinery. The complexity of this mechanism also suggests that it requires many processes to regulate it.

Complexes involved in chromatin remodeling and transcription, including `nuclear nucleosome`, `Ino80 complex`, `replication fork protection complex`, `ASTRA complex`, and `Swr1 complex`, are also highly represented. The functional cluster `vacuolar proton-transporting V-type ATPase complex` is known to have diverse roles and is associated with a wide array of processes [42].

Apart from that, we also observe the existence of several 'currency structures' – structures that may be acted upon by proteins from multiple processes. They are generally not specific to a single biological process. We classify clusters `nuclear nucleosome`, `nuclear microtubule`, `cytoplasmic microtubule`, and `extra-cellular region` as such.

Next, we analyze the BP functional cluster hubs. From Table 4.6, we found many translation related processes (`regulation of translational initiation`, `translational elongation`, `translational termination`, `tRNA aminoacy-lation` for `protein translation`, `negative regulation of translation`, `positive regulation of translation`, `ribosomal small subunit assembly`, `ribosomal large subunit assembly`). Chromatin assembly and remodeling processes (nucleosome assembly and nucleosome disassembly) also serve as key process hubs. Finally, we found major post-translation protein modification and transport processes, such as protein refolding, ATP synthesis coupled proton transport, co-translational protein targeting to membrane, and proteasome assembly, acting as hubs.

4.10 Conclusions

In this chapter, we present FUSE, a data-driven and generic algorithm for generating functional summaries at multiple resolutions from a PPI to provide a high level view of its functional landscape. It generates the "best" summary from both interaction and annotation data by maximizing information gain for a specific resolution. Our experimental study with real-world PPIs revealed that FUSE is effective and have higher accuracy compared to graph clustering techniques. It is also robust against incomplete interaction knowledge (e.g., AD network in *IntAct*). We note that the graph clustering techniques have the ability to uncover novel complexes whereas FUSE is designed to determine process-process, complex-complex, and process-complex associations with higher confidence. In this aspect, network clustering and functional summarization play complementary roles in addressing the information overload problem.

References

1. B.-S. Seah, S.S. Bhowmick, C.F. Dewey, H. Yu, FUSE: a profit maximization approach for functional summarization of biological networks. BMC Bioinf. **13**(3), S10 (2012)
2. C.H. Wu, R. Apweiler, A. Bairoch, D.A. Natale, W.C. Barker, B. Boeckmann, S. Ferro, E. Gasteiger, H. Huang, R. Lopez, M. Magrane, M.J. Martin, R. Mazumder, C. O'Donovan, N. Redaschi, B. Suzek, The universal protein resource (UniProt): an expanding universe of protein information. Nucl. Acids Res. **34**, 87–91 (2006)
3. S. Kerrien, Y. Alam-Faruque, B. Aranda, I. Bancarz, A. Bridge, C. Derow, E. Dimmer, M. Feuermann, A. Friedrichsen, R. Huntley, C. Kohler, J. Khadake, C. Leroy, A. Liban, C. Lieftink, L. Montecchi-Palazzi, S. Orchard, J. Risse, K. Robbe, B. Roechert, D. Thorneycroft, Y. Zhang, R. Apweiler, H. Hermjakob, IntAct–open source resource for molecular interaction data. Nucl. Acids Res. **35**, D561–565 (2007)
4. E.I. Boyle, S. Weng, J. Gollub, H. Jin, D. Botstein, J.M. Cherry, G. Sherlock, GO::termfinder–open source software for accessing Gene Ontology information and finding significantly enriched gene ontology terms associated with a list of genes. Bioinformatics (Oxford, England) **20**, 3710–3715 (2004)
5. G.D. Bader, C.W.V. Hogue, An automated method for finding molecular complexes in large protein interaction networks. BMC Bioinf. **4**, 2 (2003)
6. I. Dhillon, Y. Guan, B. Kulis, A fast kernel-based multilevel algorithm for graph clustering, in *Proceedings of the eleventh ACM SIGKDD international conference on knowledge discovery in data mining* (ACM, 2005), p. 634
7. C.G. Rivera, R. Vakil, J.S. Bader, NeMo: network module identification in cytoscape. BMC Bioinf. **11**(1), S61 (2010)

8. C. Kingsford, S. Navlakha, Exploring biological network dynamics with ensembles of graph partitions. in *Pacific Symposium Biocomputing* (2010), pp. 166–177
9. Y. Zhou, H. Cheng, J. Yu, Graph clustering based on structural/attribute similarities. Proc. VLDB Endow. **2**(1), 718–729 (2009)
10. N. Wang, S. Parthasarathy, K.-L. Tan, A.K.H. Tung, CSV, in *Proceedings of the 2008 ACM SIGMOD international conference on management of data - SIGMOD'08* (ACM Press, New York, USA, 2008), p. 445
11. A.-C. Gavin, M. Bösche, R. Krause, P. Grandi, M. Marzioch, A. Bauer, J. Schultz, J.M. Rick, A.-M. Michon, C.-M. Cruciat, M. Remor, C. Höfert, M. Schelder, M. Brajenovic, H. Ruffner, A. Merino, K. Klein, M. Hudak, D. Dickson, T. Rudi, V. Gnau, A. Bauch, S. Bastuck, B. Huhse, C. Leutwein, M.-A. Heurtier, R.R. Copley, A. Edelmann, E. Querfurth, V. Rybin, G. Drewes, M. Raida, T. Bouwmeester, P. Bork, B. Seraphin, B. Kuster, G. Neubauer, G. Superti-Furga, Functional organization of the yeast proteome by systematic analysis of protein complexes. Nature **415**, 141–147 (2002)
12. J. Rissanen, Modeling by shortest data description? Automatica **14**, 465–471 (1978)
13. C. Huttenhower, E.M. Haley, M.A. Hibbs, V. Dumeaux, D.R. Barrett, H.A. Coller, O.G. Troyanskaya, Exploring the human genome with functional maps exploring the human genome with functional maps. Genome Res. 1093–1106 (2009)
14. G. Palla, I. Derényi, I. Farkas, T. Vicsek, Uncovering the overlapping community structure of complex networks in nature and society. Nature **435**, 814–818 (2005)
15. B. Adamcsek, G. Palla, I.J. Farkas, I. Derényi, T. Vicsek, CFinder: locating cliques and overlapping modules in biological networks. Bioinformatics (Oxford, England) **22**, 1021–1023 (2006)
16. Y. Tian, R.A. Hankins, J.M. Patel, Efficient aggregation for graph summarization, in *Proceedings of the 2008 ACM SIGMOD international conference on management of data - SIGMOD'08* (ACM Press, New York, USA), p. 567
17. Z. Xu, Y. Ke, Y. Wang, H. Cheng, J. Cheng, A model-based approach to attributed graph clustering, in *Proceedings of the 2012 international conference on management of data - SIGMOD'12* (ACM Press, New York, USA), p. 505
18. S. Berchtold, C. Böhm, D.A. Keim, H.-P. Kriegel, A cost model for nearest neighbor search in high-dimensional data space, in *Symposium on Principles of Database Systems* (1997)
19. H. Kriegel, P. Kroger, M. Renz, S. Wurst, A generic framework for efficient subspace clustering of high-dimensional data (No. Icdm, IEEE, 2005)
20. D. Koutra, U. Kang, et al., VoG: summarizing and understanding large graphs, in *Proceedings of SDM* (2014)
21. N. Zhang, Y. Tian, J.M. Patel, Discovery-driven graph summarization, in *Proceedings of ICDE* (2010), pp. 880891
22. K. LeFevre, E. Terzi, Grass: graph structure summarization. SDM (2010), pp. 454465
23. P.K. Chan, M.D.F. Schlag, J.Y. Zien, *Spectral K -way Ratio-Cut Partitioning and Clustering* (ACM Press, New York, 1993)
24. S. Khuller, A. Moss, J.S. Naor, The budgeted maximum coverage problem. Inf. Process. Lett. **70**, (1) (1999)
25. T.S. Keshava Prasad, R. Goel, K. Kandasamy, S. Keerthikumar, S. Kumar, S. Mathivanan, D. Telikicherla, R. Raju, B. Shafreen, A. Venugopal, L. Balakrishnan, A. Marimuthu, S. Banerjee, D.S. Somanathan, A. Sebastian, S. Rani, S. Ray, C.J. Harrys Kishore, S. Kanth, M. Ahmed, M.K. Kashyap, R. Mohmood, Y.L. Ramachandra, V. Krishna, B.A. Rahiman, S. Mohan, P. Ranganathan, S. Ramabadran, R. Chaerkady, A. Pandey, Human protein reference database-2009 update. Nucl. Acids Res. **37**, D767–D772 (2009)
26. H.W. Mewes, D. Frishman, U. Güldener, G. Mannhaupt, K. Mayer, M. Mokrejs, B. Morgenstern, M. Münsterkötter, S. Rudd, B. Weil, MIPS: a database for genomes and protein sequences. Nucl. Acids Res. **30**, 31–34 (2002)
27. N.J. Krogan, G. Cagney, H. Yu, G. Zhong, X. Guo, A. Ignatchenko, J. Li, S. Pu, N. Datta, A.P. Tikuisis, T. Punna, J.M. Peregrín-Alvarez, M. Shales, X. Zhang, M. Davey, M.D. Robinson, A. Paccanaro, J.E. Bray, A. Sheung, B. Beattie, D.P. Richards, V. Canadien, A. Lalev, F. Mena,

P. Wong, A. Starostine, M.M. Canete, J. Vlasblom, S. Wu, C. Orsi, S.R. Collins, S. Chandran, R. Haw, J.J. Rilstone, K. Gandi, N.J. Thompson, G. Musso, P. St Onge, S. Ghanny, M.H.Y. Lam, G. Butland, A.M. Altaf-Ul, S. Kanaya, A. Shilatifard, E. O'Shea, J.S. Weissman, C.J. Ingles, T.R. Hughes, J. Parkinson, M. Gerstein, S.J. Wodak, A. Emili, J.F. Greenblatt, Global landscape of protein complexes in the yeast Saccharomyces cerevisiae. Nature **440**, 637–643 (2006)

28. D. Crabtree, P. Andreae, X. Gao, QC4: a clustering evaluation method, in *Proceedings of the 11th Pacific-Asia Conference on Advances in Knowledge Discovery and Data Mining (PAKDD'07)* (2007), pp. 59–70
29. M. Koyutürk, W. Szpankowski, A. Grama, Assessing significance of connectivity and conservation in protein interaction networks. J. Comput. Biol. **14**(6), 747–764 (2007)
30. A. Ben-Hur, A. Elisseeff, I. Guyon, A stability based method for discovering structure in clustered data, in *Biocomputing 2002 - Proceedings of the Pacific Symposium* (World Scientific Publishing Co. Pte. Ltd., Singapore, 2001), pp. 6–17
31. K.J. De Vos, A.J. Grierson, S. Ackerley, C.C.J. Miller, Role of axonal transport in neurodegenerative diseases. Annu. Rev. Neurosci. **31**, 151–173 (2008)
32. D.J. Owen, B.M. Collins, Vesicle transport: a new player in APP trafficking. Curr. Biol. **20**, R413–R415 (2010)
33. M.T. Lin, M.F. Beal, Mitochondrial dysfunction and oxidative stress in neurodegenerative diseases. Nature **443**, 787–795 (2006)
34. D.J. Selkoe, Folding proteins in fatal ways. Nature **426**, 900–904 (2003)
35. K.V. Kuchibhotla, S.T. Goldman, C.R. Lattarulo, H.-Y. Wu, B.T. Hyman, B.J. Bacskai, Abeta plaques lead to aberrant regulation of calcium homeostasis in vivo resulting in structural and functional disruption of neuronal networks. Neuron **59**, 214–225 (2008)
36. K. Herrup, Y. Yang, Cell cycle regulation in the postmitotic neuron: oxymoron or new biology? Nat. Rev. Neurosci. **8**, 368–378 (2007)
37. R.A.C.M. Boonen, P. van Tijn, D. Zivkovic, Wnt signaling in Alzheimer's disease: up or down, that is the question. *Ageing Res. Rev.* **8**, 71–82 (2009)
38. B.V. Zlokovic, Neurovascular mechanisms of Alzheimer's neurodegeneration. Trends Neurosci. **28**, 202–208 (2005)
39. B.W. Doble, GSK-3: tricks of the trade for a multi-tasking kinase. J. Cell Sci. **116**, 1175–1186 (2003)
40. B. De Strooper, W. Annaert, Where notch and wnt signaling meet. the presenilin hub. J. Cell Biol. **152**, F17–F20 (2001)
41. A. Patil, H. Nakamura, Disordered domains and high surface charge confer hubs with the ability to interact with multiple proteins in interaction networks. FEBS Lett. **580**, 2041–2045 (2006)
42. N. Nelson, N. Perzov, A. Cohen, K. Hagai, V. Padler, H. Nelson, The cellular biology of proton-motive force generation by V-ATPases. J. Exp. Biol. **203**, 89–95 (2000)

Chapter 5
Multi-faceted Functional Decomposition

In this chapter, we present a PPI decomposition algorithm called FACETS [1] in order to make sense of the deluge of interaction data using GO annotation data. A key distinguishing feature of FACETS is that it finds not just a single functional decomposition of the PPI network, but a *multi-faceted atlas* of functional decompositions that portray alternative perspectives of the functional landscape of the underlying PPI. Each *facet* in the atlas represents a distinct interpretation of how the network can be functionally decomposed and organized. Specifically, the FACETS algorithm maximizes interpretative value of the atlas by optimizing *inter-facet orthogonality* and *intra-facet cluster modularity*.

5.1 Motivation

Recall that graph clustering algorithms [2–4] discover regions of dense connectivity that represent protein complexes or functionally coherent processes. Unfortunately, these methods output *only a single optimal functional decomposition* of the PPI network. Consequently, a PPI network can only be decomposed and viewed from a single perspective, whereas in reality there are often multiple different perspectives (decompositions) associated with the functional organization of the underlying network, all of which are distinct and equally valid. We refer to each of these decompositions as a *facet* because they visualize the organization of a PPI network from a unique view, providing a distinct interpretation of the organization of the underlying network. For example, consider the toy transcriptional regulatory network depicted in Fig. 5.1. A typical decomposition, based on an existing graph clustering technique (e.g., MCODE in [2]), identifies dense regions of the network, which correspond to the decomposition of protein complexes as shown in *Facet 1*. However, this network can also be viewed from other different perspectives. For instance, it can be organized by the types of signaling pathways involved in it (*Facet 2*). Notice that the decomposition

S.S. Bhowmick and B.-S. Seah, *Summarizing Biological Networks*,
Computational Biology 24, DOI 10.1007/978-3-319-54621-6_5

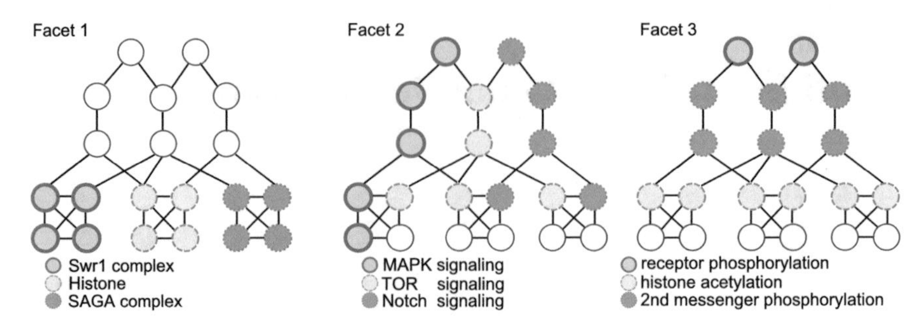

Fig. 5.1 Illustration of multi-faceted PPI network decomposition

from this perspective is markedly different from the complex-based decomposition. Furthermore, different proteins in the network may undergo various modifications such as acetylation, phosphorylation, and ubiquitination. Hence, yet another way to decompose the network is by their modification effects as depicted in *Facet 3*. Clearly, in larger real-world networks the possibility of uncovering multiple, distinct functional decompositions is real.

At first glance, it may seem that we can tune the clustering parameters of existing graph clustering techniques in order to generate multiple facets or decompositions. Unfortunately, such tuning only generates an exponential number of *slightly perturbed* decompositions with incremental differences. In other words, such strategy does not generate functionally unique decompositions. In contrast, it is imperative to ensure that the decompositions or facets are *distinctive*, i.e., they are maximally different from each other. This is because every facet should provide a fresh and informative perspective to the organization of the network, rather than providing just incremental differences with respect to other facets.

This chapter presents an algorithm called FACETS [1] that discovers an *atlas* of functionally unique decompositions from a PPI network, portraying alternative views of the functional landscape of the network (detailed in Sect. 5.3). Each decomposition or facet represents a distinct interpretation of how the network can be functionally decomposed and organized. Since a key objective is to obtain n unique facets that are informative and orthogonal,[1] the algorithm maximizes interpretative value of the atlas by optimizing *intra-facet cluster modularity* and *inter-facet orthogonality*. *Intra-facet cluster modularity* captures the aim of decomposing a PPI network G based on a particular functional and/or structural view. For instance, based on complexes and localized structures, G can be decomposed into protein complexes. If we consider regulatory processes as a functional concept, then G can be decomposed into signaling and regulation pathways, an entirely different decomposition. *Inter-facet orthogonality*, on the other hand, demands that the n facets are structurally distinctive and functionally apart from each other. Hence an *objective function* is proposed that models these intuitions and FACETS exploit it to discover a set of distinct facets.

[1]We use the term orthogonal to describe the idea of distinctive clusters, rather than its precise mathematical meaning.

Specifically, both the PPI graph structure and the rich functional information provided by GO annotations are exploited to guide facets construction.

5.2 Related Work

Multi-view clustering is a poorly studied problem in the data mining community [5]. Still, there are several work that have focused on multi-view clusterings in image and text mining domain [6]. One approach projects data into an alternative subspace [7]. Another approach generates alternative clustering through the use of *must-link* and *cannot-link* constraints [8]. In Meta-Clustering [9], a large number of clusters are generated, and clusters which are truly different are selected. All of the aforementioned approaches, however, assume data points in the *vector space* that allow the notion of metric distances in a Euclidean geometry. On the other hand, our problem demands a multi-view clustering methodology on *attributed graphs*, which requires a graph clustering paradigm on both structure and annotation. To the best of our knowledge, multi-view clustering paradigm has not been applied in clustering biological networks to identify pertinent functional modules from multiple perspectives.

Ensemble clustering methods generate an ensemble of near-optimal decompositions [10–12]. These methods have been used to increase the quality and confidence of the decomposition and understand network dynamics. The near-optimal decompositions generated, however, have no notion of orthogonality that this work is seeking. Instead, ensemble clusterings create a large number of perturbed solutions, making them unsuitable as an atlas of functionally distinct decompositions. For instance, in [13], a small network of 32 nodes generated at least 82 permutations of clusterings.

5.3 Problem Statement

In this section, we formally introduce the multi-faceted functional decomposition problem. We begin by defining some terminology that we shall be using in the subsequent discussion in this chapter. We use the network in Fig. 5.1 as running example.

5.3.1 *Terminology*

A *facet* (*decomposition* or *view*) of G, denoted by F, is a set of functional modules $\{C_1, \ldots, C_m\}$ representing a specific functional concept. Functional modules within a facet F are allowed to overlap. In the sequel, the terms facet, view, and decomposition are used interchangeably. A *functional atlas* (or atlas for brevity) of G, denoted by A, is a set of facets $\{F_1, F_2, \ldots, F_n\}$ that represents distinctive functional landscapes of G. Figure 5.1 depicts an atlas of 3 facets, with each facet decomposing the network into 3 functional modules.

Similar to FUSE, GO annotations associated with proteins are utilized to define the multi-faceted functional decomposition problem. Given a GO directed acyclic graph (DAG) $D = (V_{go}, E_{go})$, the ordered set $\Delta = \langle \Delta_1, \Delta_2, \ldots, \Delta_d \rangle$ is a topological sort of D, where Δ_i represents a single GO term. Each vertex $v \in V$ is associated with a d-dimensional *function association vector* $\Delta_v \in \{0, 1\}^d$, such that $\Delta_v = \langle \Delta_1^v, \Delta_2^v, \ldots, \Delta_d^v \rangle$, $\Delta_i^v \in \{0, 1\}$ where $\Delta_i^v = 1$ if and only if the term $\Delta_i \in D$ or its descendants are associated with protein v, and $\Delta_i^v = 0$ if otherwise.

A *facet candidate bundle* $B_i = \{G_1, G_2, \ldots, G_m\}$ is a set of connected subnetworks of G such that for every $G_k \in B_i$, there is a shared GO term Δ_i within every $v \in V_k$. Δ_i represents the common function of the candidate subnetwork. A facet candidate bundle B_i represents the superset of facet F_i, and it contains a large permutation of subnetworks that satisfy a particular functional concept. Typically, $|F_i| \ll |B_i|$. A *function bundle* $\omega_i = \{\Delta_1, \Delta_2, \ldots \Delta_m\}$ is the set of shared GO annotations of B_i, i.e., $\omega_i = \bigcup_{G_k \in B_i} \Delta_{G_k}$. To illustrate these concepts, consider the PPI network in Fig. 5.1. Suppose that B_1 is a facet candidate bundle with $\omega_1 = \{\Delta_1, \Delta_2\}$, where Δ_1 represents the `Swr1 complex` GO term and Δ_2 the `Histone` term. In the subgraph with '`Swr1 complex`' label in Facet 1, every node in that subgraph is annotated with `Swr1 complex` term. Thus, the subgraph is a valid member of B_1. Any subgraph made up of '`Swr1 complex`'-labeled nodes is also a valid member of B_1. If B_2 represents the facet candidate bundle with $\omega_2 = \{\Delta_3\}$, where Δ_3 represents `cellular component`, then the '`Swr1 complex`'-labeled subgraph is also a valid member of B_2 (`Swr1 complex` is a `cellular component`). Furthermore, every subgraph in Facet 1 whose nodes are labeled is a valid member of B_2, but not necessarily a valid member of B_1. One can see that B_i contains a set of subgraphs that shares specific functional concepts depending on the functional terms in ω_i. We define the function $f: \mathscr{P}(V_{go}) \rightarrow A$ given by $f(\omega_i) = F_i$ to make explicit the association between a functional bundle and its corresponding facet.

A *function bundle partition* $\Omega = \{\omega_1, \omega_2, \ldots, \omega_n\}$ is a set of function bundles that forms a partition of all GO terms V_{go}, i.e., $\bigcup_{\omega_i \in \Omega} = V_{go}$. In the next section, further constraints on facet candidate bundles and function bundles are imposed such that the shared GO terms of the subnetworks within each facet candidate bundle share high functional commonality and the terms share in one facet are distinct from the terms in another facet.

5.3.2 Multi-faceted Functional Decomposition Problem

The goal of multi-faceted functional decomposition problem is to identify an atlas of n distinct facets of G that maximizes *inter-facet orthogonality* and *intra-facet cluster modularity*. Each facet depicts a higher-order organization of modules of G. Recall that inter-facet functional orthogonality demands that each of the n facets is based on an orthogonal functional concept – facets that are distinctive and functionally apart from each other. Hence, two criteria are proposed that model the intra-facet functional modularity and inter-facet orthogonality of an atlas solution. Next, an *objective function* is introduced that models and scores an atlas of G.

Intra-facet cluster modularity. Intra-facet cluster modularity enables us to seek clusters that are both structurally and functionally modular. Given ω_i, Ω, and G, *ω-restricted* decomposition procedure (denoted by g_ω) computes a decomposition of G into F_i such that F_i satisfies the following criteria:

- *Criterion 1.* Every module $C_j \in F_i$ should be *functionally bounded* by ω_i. Let $D_{C_j} = \{\Delta_1, \Delta_2, \ldots, \Delta_m\}$ be the set of shared terms in C_j, i.e., for every $v \in V_c^j$, v must be annotated with every $\Delta_i \in D_{C_j}$. Then, the *functional boundedness* of module C_j by ω_i is given by $r(C_j, \omega_i) = D_{C_j} \cap \omega_i$. A cluster C_j is bounded by ω_i if $r(C_j, \omega_i) \neq \emptyset$. An ω_i-restricted decomposition of a facet draws from a restricted search space of subnetworks in G whose vertices share at least a term within ω_i. Intuitively, this means that for any subnetwork to be considered as a module, it must first be sharing a term in ω_i. Even if a subnetwork is dense, it must yield to sparser subnetwork candidates if it is not enriched with terms within ω_i. In the example of Fig. 5.1, if ω_1 is terms of protein complexes, then any subgraphs enriched with complex terms is in the search space for *Facet 1*. In contrast, the modules of *Facet 2*, enriched with signaling terms, would be invalid candidates for *Facet 1* decomposition. This restricted search space is modeled by facet bundle B_i, where any valid candidate facet cluster C_j of facet F_i must belong to B_i.
- *Criterion 2.* A facet F_i decomposes G by maximizing a clustering objective function $o(F_i)$ while satisfying the above criterion. $o(F_i)$ is determined by the specific graph clustering algorithm that is adapted for creating a facet; for generality we let this be the objective function $o(F_i)$ that has to be maximized by the graph clustering algorithm. For instance, every module $C_j \in F_i$ has to be structurally dense and/or functionally coherent (i.e., every node in a module shares a common function), the coverage of F_i has to be high, and the amount of overlap between modules should be low. For example, modules of *Facet 2* maximize $o(F_2)$ while satisfying the ω_2 bound, despite not forming dense modules. This is because all dense modules formed are enriched with complex terms, violating the ω_2 bound.

Inter-facet orthogonality. Since we want every facet in the atlas to be functionally and structurally distinct, modules within a facet, as whole, should be structurally and functionally distinct from modules within another facet. We discuss two independent distance measure between facets: *functional orthogonality* and *structural orthogonality*.

Functional orthogonality is indirectly controllable by the function bundles attached to facets, which determines the types of allowable modules through the aforementioned restriction. By increasing inter-bundle functional orthogonality, we increase the functional distinctiveness of each facet. To impose functional orthogonality, we introduce the following constraint: for every $\omega_i, \omega_j \in \Omega$, $\omega_i \cap \omega_j = \emptyset$ and $\bigcup_{\omega_i \in \Omega} = V_{go}$. This requires that Ω actually partitions the terms of the GO DAG. The *functional distance measure* between Δ_i and Δ_j, denoted by $d(\Delta_i, \Delta_j)$, measures the functional dissimilarity between the terms. Here, $d(\Delta_i, \Delta_j)$ is simply computed as the length of the shortest path between the terms: $d_f(\Delta_i, \Delta_j) = min_{\Delta_r \in R}|p(\Delta_r, \Delta_i)| + |p(\Delta_r, \Delta_j)|$, where R is the set of common ancestors of term Δ_i and Δ_j and $|p(i, j)|$ is the length of the shortest path from node Δ_i to Δ_j in D. The *candidate function specificity* $s(\Delta_i, C_u)$ is defined as

$$s(\Delta_i, C_u) = \frac{|\{\Delta_i \in \Delta_v | v \in V_c^u\}|}{|\{\Delta_i \in \Delta_v | v \in V\}|}$$

In the above equation, $s(\Delta_i, C_u)$ measures the specificity of a shared GO term, which we will later use to weigh the contribution of the term. For instance, a cluster C_j of 5 nodes that share the `biological process` GO term in a network of 1000 `biological process` annotated nodes has a low specificity value of 0.005 with respect to the term.

Likewise, we define structural orthogonality. The *structural distance measure* between two clusters C_u and C_v is defined as

$$d_s(C_u, C_v) = 1 - |E_C^u \cap E_C^v| / |\{(v_i, v_j) | v_i \in V_C^u \cap V_C^v, v_j \in V_C^u \cup V_C^v, (v_i, v_j) \in E_C^u \cup E_C^v\}| \quad (5.1)$$

$d_s(C_u, C_v)$ measures difference between 1 and the ratio of the number shared edges between C_u and C_v over the number of edges incident to $V_C^u \cap V_C^v$. The distance is 0 if C_u and C_v shares all edges and 1 if C_u and C_v shares no edges.

Following that, we define $t(\Omega, A)$ as the linear combination of inter-facet functional and structural orthogonality, as follows:

$$\begin{aligned} t(\Omega, A) = \gamma \sum_{\substack{\omega_i, \omega_j \in \Omega \\ i \neq j}} \{ \sum_{\substack{\Delta_j \in D_{C_j} \\ C_j \in f(\omega_j)}} \sum_{\substack{\Delta_i \in D_{C_i} \\ C_i \in f(\omega_i)}} s(\Delta_i, C_i) s(\Delta_j, C_j) \frac{d_f(\Delta_j, \Delta_i)}{|V_p^j||V_p^i|} \} \\ + (1 - \gamma) \sum_{\substack{\Delta_u \in D_{C_u}, C_u \in F_i \\ F_i \in A}} \sum_{\substack{\Delta_v \in D_{C_v}, C_v \in F_j \\ F_j \in A, i \neq j}} s(\Delta_u, C_u) s(\Delta_v, C_v) d_s(C_u, C_v) \end{aligned} \quad (5.2)$$

The parameter γ weighs the contribution of d_s against d_f, and is set to attain balanced contribution from both terms. Note that $t(\Omega, A)$ quantifies the pairwise orthogonality between two function bundles. The higher the score, the greater the orthogonality.

5.3.3 Problem Definition

The multi-faceted functional decomposition of G is defined as the problem of simultaneously constructing the atlas of decompositions $A = \{F_1, \ldots, F_n\}$, and the function partition $\Omega = \{\omega_1, \ldots, \omega_n\}$, such that the following objective function is maximized:

$$\begin{aligned} \max_{A,\Omega} \quad & \lambda t(\Omega, A) + (1 - \lambda)|A|^{-1} \sum_{F_i \in A} o(F_i) \\ \text{subject to} \quad & C_s \in B_i \forall C_s \in F_i, 1 \leq i \leq n \end{aligned} \tag{5.3}$$

The right half of the terms captures the cost function of decomposing G into A; the left half, decomposing D into Ω. The parameter $\lambda \in [0, 1]$ controls the weight between the two terms. Observe that one has to optimize these criteria simultaneously over the space of A and Ω. Otherwise, one may end up with a poor objective score. For instance, if $t(\Omega, A)$ is high (meaning highly orthogonal partitioning), but Ω is improperly partitioned such that one ends up with ω_i that allow only poor decompositions, then the $o(F_i)$ score would be very low. Due to the interdependence of the criteria, optimizing the aforementioned function is computationally expensive.

5.4 FACETS Algorithm

Generally, the problem of finding clusters that maximizes typical clustering objective functions that relate to graph density is known to be NP-hard [14]. Hence the FACETS algorithm is a heuristic implementation that attempts to find a local maximum of the objective function. The heuristic deployed is a step-wise iterative approach that incrementally optimizes Ω and A, one at a time. Intuitively, given an attributed PPI network (e.g., Fig. 5.2a), Ω is incrementally updated by using each facet in A as functional centroids, and then using the centroids to partition D. A is updated through ω-restricted decomposition using the updated Ω. The FACETS algorithm consists of two phases: the *initialization* phase (Fig. 5.2b), and the *iteration* phase (Fig. 5.2c, d). We describe each of them in turn.

5.4.1 The Initialization Phase

This phase creates an initial set of decompositions for the second phase. It performs graph clustering on G to obtain an initial set of modules. To this end, the FUSE algorithm is utilized. Each module of this set is then randomly associated with a facet, randomly distributing the modules over an initial set of facets. Following that, we construct *candidates subnetworks*, which use subnetworks of G that sat-

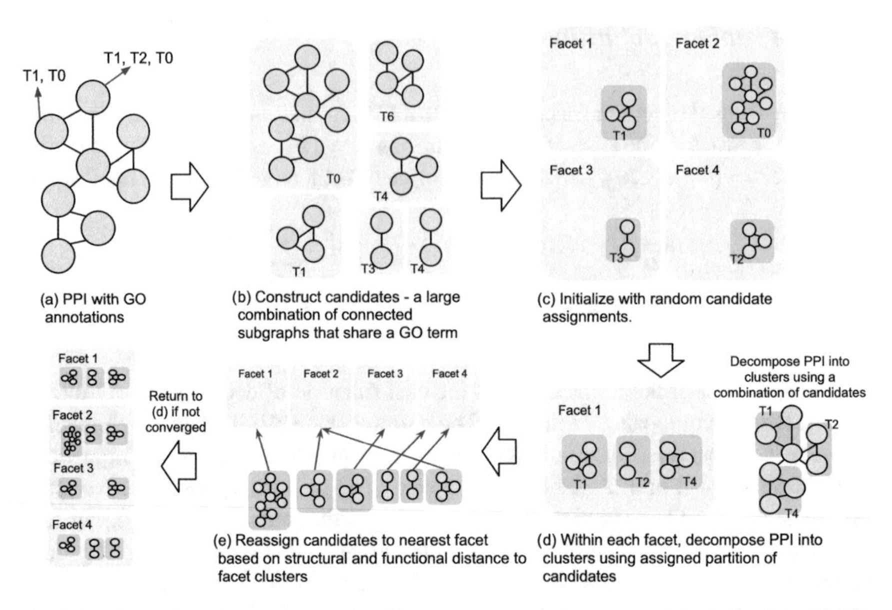

Fig. 5.2 Illustration of the FACETS algorithm. **a** GO annotated PPI network is used as input. **b** The set of candidate subnetworks are computed **c** An initial set of modules are randomly assigned to a facet. Candidate subnetworks are then assigned to their nearest facet based on function and structure distance. **d** For each facet, decomposition is performed to identify modules that are functionally contained by the facet candidate bundle. **e** The candidate subnetworks are reassigned based on their distance to the new set of modules identified. Convergence is achieved when the number of terms reassigned to a different facet drops below the threshold parameter θ. Otherwise, steps d–e are repeated

isfy ω_i-restricted decomposition constraint. To generate candidates exhaustively is prohibitively expensive. Instead, candidates for a facet F_i are generated as follows: for every GO term $\Delta \in \omega_i$, we obtain the induced subnetwork in G whose nodes are annotated with Δ or its descendants. The subnetwork is then decomposed into connected components, each forming a candidate subnetwork G_j. Let $\Delta_j^C = \Delta$ be the term associated with this candidate. Candidates formed this way can vary greatly in resolution of the annotation that its nodes share (for example, $\Delta_j^C =$ `biological process`), and can be highly overlapping.

5.4.2 The Iteration Phase

This phase – the actual optimization phase – is performed in rounds. Let i denote the i-th iteration of the algorithm. At each round, the algorithm updates A and Ω in two sequential steps. To evaluate algorithm convergence, we introduce *functional reassignment* – the number of terms in Δ that is reassigned to a different function bundle after step 1 of i-th iteration. This score measures the rate of change of Ω, indicating how close the algorithm is to convergence. Observe that when Ω is fixed, the algorithm reaches a steady state. The algorithm reaches convergence and terminates when the functional reassignment at i-th iteration drops below *convergence threshold* θ, a user-defined parameter.

Update Ω. In this step, we assume that A is a constant and update Ω to increase $t(\Omega, A)$. For each $F_i \in A$, the enriched functional terms of the modules in F_i serve as centroids for partitioning D into orthogonal concepts; these enriched terms as whole form the centroid of ω_i, which is associated with F_i. We then reassign every candidate subnetwork to its nearest centroid to form a partition Ω. The convergence properties of such centroid-based partitioning approaches (e.g., K-Means) has been well studied [15]. For every $G_j \in B_i$, $1 \leq i \leq n$, we determine its *closest* centroid by considering G_j's average functional and structural distance to functional modules within a facet. The facet that is closest to G_j is indicated by:

$$d_c(G_j, F_k) = \begin{cases} 1 \text{ if} \dfrac{1}{Z(F_k)} \displaystyle\sum_{C_i \in F_k} s(\Delta_i^C, C_i)\phi(C_i, G_j) \\ \leq \dfrac{1}{Z(F_{k'})} \displaystyle\sum_{C_i \in F_{k'}} s(\Delta_i^C, C_i)\phi(C_i, G_j) \\ k' \neq k, \text{where} \\ \phi(C_i, C_j) = \gamma \dfrac{d_f(\Delta_i^C, \Delta_j^C)}{|V_p^j||V_p^i|} + (1-\gamma) d_s(C_i, C_j) \\ Z(F) = \displaystyle\sum_{C_i \in F} s(\Delta_i^C, C_i) \\ 0 \text{ otherwise} \end{cases} \tag{5.4}$$

Following that, G_j is reassigned to nearest facet candidate bundle B_k (superset of F_k) and Ω is updated based on where every $\Delta_j^C \in V_{go}$ is assigned to. Each function bundle $\omega_i \in \Omega$ represents functional terms that are most closely associated with F_i, and the decomposition of F_i in the following step will be bounded by the updated ω_i. Function partitioning depends on the atlas of decompositions because not every partition of the GO DAG is capable of forming a modular decomposition of functional modules.

Update A. In this step, we update A to maximize the objective function while fixing Ω. To support ω_i-restricted decomposition of F_i, we propose an algorithm that employs profit maximization model (discussed below) and runs in iterations. At each iteration, we score candidate subnetworks based on a profit maximization model and greedily selects the best scoring candidate as member in F_i. An iteration runs for every $F_i \in A$ before moving to the next iteration. Every candidate considered for F_i must satisfy the ω_i-restricted decomposition constraint, i.e., the candidate subnetwork must be enriched with terms in ω_i. In other words, $G_j \in B_i$.

We now describe the profit maximization model for scoring a candidate $G_j \in B_i$. Every $v \in V$ is assigned an information budget. A candidate G_j extracts, from each $v \in V_j^G$, some information revenue from the budget pool. The revenue extracted is

Algorithm 3 Algorithm FACETS

Input: $G, \Delta, D, k, b, d, \beta, n, \theta$
Output: A

1: $S = \text{FUSE}(G, \Delta, D, k, b, d, \beta)$
2: $A = \{F_1, F_2, \ldots, F_n\}$ where $F_i = \emptyset$
3: **for** $C(u) \in S$ **do**
4: $\quad F_r = F_r \cup C(u)$ where r is randomly 1 to n
5: **end for**
6: **while** $reassignment(\Omega_{old}, \Omega) > \theta$ **do**
7: $\quad \Omega_{old} = \Omega$
8: $\quad \Omega = \{B_1, B_2, \ldots, B_n\}$ where $B_i = \emptyset$
9: $\quad$ **for** $F_i \in A$ **do**
10: $\quad\quad$ **for** $G_j \in B_i, B_i \in \Omega_{old}$ **do**
11: $\quad\quad\quad F_{min} = argmin_{F_i} d_c(F_i, G_j)$
12: $\quad\quad\quad B_{min} = B_{min} \cup G_j$
13: $\quad\quad$ **end for**
14: $\quad$ **end for**
15: $\quad \Omega = \{B_1, B_2, \ldots, B_n\}$
16: $\quad$ **for** $F_i \in A$ **do**
17: $\quad\quad F_i = \text{FUSE}(G, \Delta, B_i, k, b, d, \beta)$
18: $\quad$ **end for**
19: **end while**
20: **for** $F_i \in A$ **do**
21: $\quad$ **for** $C(i), C(j) \in F_i$ **do**
22: $\quad\quad$ **if** $C(i) \neq C(j)$ and $P_i(X > occ_{C(i)C(j)}) \leq 2\beta/|S|^2$ **then**
23: $\quad\quad\quad$ Add edge $(C(i), C(j))$ to F
24: $\quad\quad$ **end if**
25: $\quad$ **end for**
26: **end for**

correlated to the edge density of the subnetwork, with modular candidates giving high revenue. Each time a candidate is selected, revenue is removed from the budget pool and a cost is incurred. A penalty cost is incurred for a candidate that is structurally similar to selected clusters in other facets $F_{i'} \neq F_i$. This penalty is modeled by $cost(G_j) = \sum_{C' \in F_{i'}, i' \neq i} d_s(G_j, C')$, which utilizes the structural distance measure d_s described earlier. At each iteration, the candidate that contributes the highest information profit (revenue minus cost) is selected. To summarize, a clustering in F_i that yields high overall revenue have subgraphs with high facet modularity $o(F_i)$, while a clustering with low overall cost yields high inter-facet orthogonality $t(\Omega, A)$. Given a fixed Ω, the set of facets A with maximum overall profit maximizes the objective function. The algorithm above approximates this through greedy heuristic.

Algorithm 3 shows the steps of the FACETS algorithm. We illustrate the algorithm with the example shown in Fig. 5.2. The initialization step constructs a set of initial candidate subnetworks and assigns them randomly to a facet (lines 1–5, Fig. 5.2b, c). Following that, the update Ω and update A steps are performed iteratively until convergence (lines 6–19, Fig. 5.2d, e). In the update Ω step, each candidate subnetwork is assigned to its nearest facet (lines 7–15, Fig. 5.2d), while in the update A, a

Table 5.1 Datasets used in FACETS

Dataset	#nodes	#edges	Source
H. sapiens	9131	34362	IntAct [16]
S. cerevisiae	4768	40457	IntAct
D. melanogaster	3114	6472	IntAct
Human autophagy	1241	3555	IntAct

restricted FUSE profit maximization heuristic is performed to identify the best set of subnetworks that represent a facet (lines 6–19, Fig. 5.2d). Finally, upon convergence, the network for each facet is constructed (lines 20–26, Fig. 5.2e).

5.5 Experimental Study

The FACETS algorithm is implemented in Scala. We now present the experiments conducted to study the performance of FACETS and report some of the results here. All experiments were executed on a 1.66 GHz Intel Core 2 Duo T5450 machine with 3 GB memory.

5.5.1 *Experiment Settings*

We primarily used the global human PPI network from *IntAct* [16], as well as the *yeast*, *fruit fly*, and *human autophagy* networks from *IntAct* (Table 5.1). In all experiments, we set the convergence threshold $\theta = 5$. The weight γ is set to 0.091 to balance the contribution of structure and function (equal order of magnitude). We utilize only the `Cellular Process` sub-domain of the Gene Ontology so that the facets are created not merely based on different GO domains, but created based on more subtle functional differences.

To measure the similarity/dissimilarity between facets or decompositions, we employed the *Jaccard index* (JI) [17] evaluation measure, which is widely used to compare clusterings based on counting the agreement or disagreement of co-clustered pairs of proteins. Given two decompositions (or facets) f_1 and f_2, the *Jaccard index* is defined as

$$J(f_1, f_2) = \frac{A}{A + B + C}$$

where A is the number of protein pairs that is co-clustered in both f_1 and f_2, B is the number of protein pairs co-clustered in f_1 but not f_2, and C is the number of protein pairs co-clustered in f_2 but not f_1. $J(f_1, f_2)$ ranges from 0 to 1 (for identical clusterings).

Table 5.2 Comparison between facets of the *H. sapiens* PPI network ($n = 6$)

			JI score					
Facet	#Modules	Coverage	Fct 1	Fct 2	Fct 3	Fct 4	Fct 5	Fct 6
1	89	294	1.0	0.014	0.065	0.0050	0.0070	0.079
2	280	1079	0.014	1.0	0.0040	0.119	0.0050	0.0070
3	106	372	0.065	0.0040	1.0	0.0010	0.0	0.013
4	94	419	0.0050	0.119	0.0010	1.0	0.0	0.0080
5	114	390	0.0070	0.0050	0.0	0.0	1.0	0.0010
6	72	306	0.079	0.0070	0.013	0.0080	0.0010	1.0
			Coverage overlap					
Facet			Fct 1	Fct 2	Fct 3	Fct 4	Fct 5	Fct 6
1			1.0	0.316	0.142	0.081	0.044	0.112
2			0.086	1.0	0.077	0.09	0.082	0.079
3			0.112	0.225	1.0	0.029	0.059	0.086
4			0.057	0.233	0.026	1.0	0.028	0.052
5			0.033	0.228	0.056	0.03	1.0	0.038
6			0.107	0.281	0.104	0.071	0.049	1.0

5.5.2 *Results*

Quantitative Assessment

Table 5.2 shows the quantitative comparison between facets. We measure the inter-facet decomposition similarity using the JI score. The low clustering similarity scores between facets show that they are decomposed distinctively. This reflects significant organizational differences between modules of signaling pathways and protein complexes. We measure the *coverage* of a facet and the *extent* of coverage overlap between the facets. Let the coverage of a facet F_k be $Cvg(F_k) = |\bigcup_{V_c \in F_k} V_c|$. Also, let the extent of coverage overlap between F_i and F_j be $Ext(F_i, F_j) = \frac{|V_i \cap V_j|}{|V_i|}$, where $V_i = \bigcup_{V_c \in F_i} V_c$ and $V_j = \bigcup_{V_c \in F_j} V_c$. The extent of overlap between facets reaches up to 0.316. Consequently, the overlap is not insignificant, implying that the facets are not partitions of G.

Validation on Real Data

In this experiment, we compare the `FACETS` atlases of the global human network to gold standard functional modules. The gold standard datasets were constructed as follows: (1) `mips` – We use the set of 571 human complexes (of more than 3 proteins) from MIPS [18] to represent the decomposition of the human interactome into complexes. (2) `kegg-metabolic` – To represent decomposition into metabolic modules, we use 67 human metabolic networks from `KEGG`, each representing a single functional module. (3) `kegg-signaling` – We use 23 human signal transduction pathways from `KEGG` to represent decomposition into signaling pathways. The gold standard decompositions were chosen such that each represents a distinct

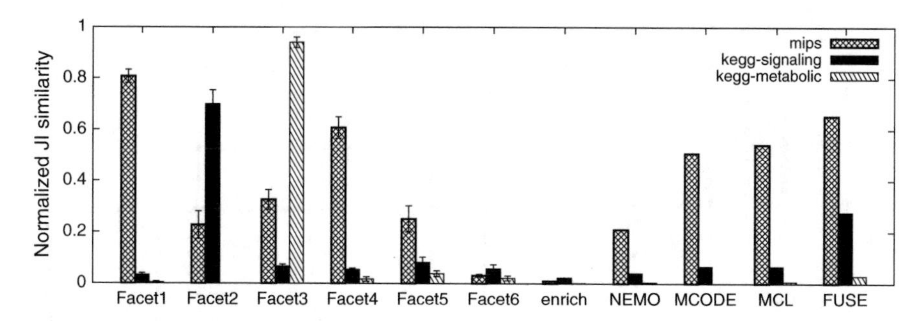

Fig. 5.3 Comparison between the decomposition similarities of FACETS, other methods, and gold standard decompositions

functional organization of the human network. As such, we consider each gold standard dataset as a facet of the human network, and the set of these three datasets as the gold standard atlas of the human network. We then compare these datasets against the atlas of facets obtained through our algorithm and determine if there is a distinctive one-to-one mapping between our facet and a gold standard facet. We set $n = 6$ and repeated the tests fifteen times under different starting conditions to account for variability in FACETS output. We also compare the similarity scores against graph clustering methods, namely Markov clustering (MCL) [3], MCODE [2], NeMo [19], and FUSE [4]. These methods create a single decomposition of the human network. We removed clusters with fewer than 3 proteins. We also compare against GO term enrichment (enrich) [20], which does not utilize structural information. Following that, we measure the clustering similarities between the gold standard datasets and the decompositions obtained. Figure 5.3 shows the clustering similarities between modules in gold standard datasets and modules in facets as well as tested graph clustering methods. The Jaccard index were used to measure the agreement between pairs of decompositions. We normalize the scores so that the highest JI score obtained, within each gold standard dataset, is adjusted to 1.

We consider the facet best associated with a gold standard decomposition by comparing their relative scores. The gold standard datasets are uniquely mapped to a distinct facet: `kegg-metabolic` is most similar to *Facet 3*, `kegg-signaling` is most similar to *Facet 2*, and `mips` is most similar to *Facet 1*. This unique mapping demonstrates that from a clustering perspective, the facets have significant functional orthogonality such that they are uniquely associated with different functional organization maps. *Facet 6* has poor similarity to the gold datasets, indicating a set of clusters that could be functionally distinct from these datasets.

In contrast, the tested graph clustering methods share common similarity patterns. Clusters are largely from a single dominant perspective – those of protein complexes (`mips`). We argue that objective functions based on dense connectivity tend to favor protein complex structures over other decompositions like metabolic pathways. GO term enrichment, on the other hand, generates output with little similarity to all gold standard datasets, indicating that annotations alone are unable to specifically identify

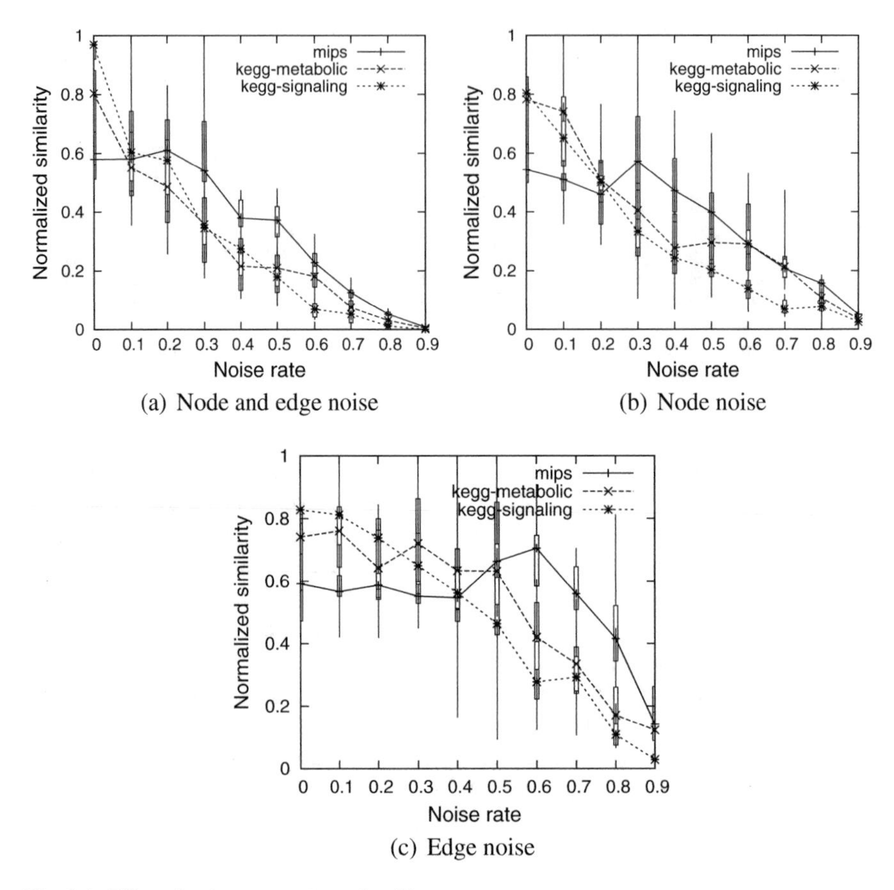

Fig. 5.4 Effect of noise on FACETS algorithm

important functional modules within a large PPI network. This is supported by the fact that functional analysis of large networks often involve graph clustering prior to term enrichment [3].

Robustness. To study the robustness of FACETS, we test the effect of annotation perturbations and edge deletions of the input network on FACETS output. Random edge deletion (*edge noise*) simulates the effect of removing false positive interactions in high-throughput interaction datasets, while annotation perturbation (*node noise*) simulates errors in curated annotations. Figures 5.4a–c show the effect of edge and node noise on FACETS, varying from 0% noise to 100% noise. The figures show clustering similarities (JI similarity) between the best scoring facets and gold standard datasets under increasing noise perturbations. We repeat each test fifteen times with different randomization seed. We observe that FACETS output quality drops gradually under increasing edge and node noise conditions. This demonstrates that the algorithm is robust to small noise perturbations. In case of edge noise, we note that the

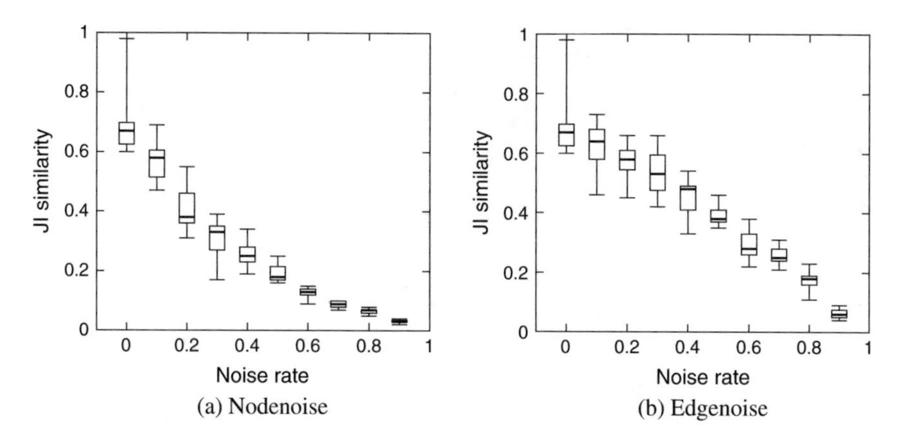

(a) Nodenoise
(b) Edgenoise

Fig. 5.5 Effect of initial starting point versus noise on FACETS algorithm

quality of output only drops rapidly past the 0.5 noise ratio. This is desirable given that false positive rates in yeast two-hybrid and TAP experiments range between 0.35 to 0.7 [21]. MIPS clusters, which consist densely interconnected clusters, are most robust to edge noise effects. The effect of node noise is comparatively greater, but quality degradation remains gradual.

Effect of initial starting point. Given that FACETS belongs to the class of hill-climbing methods, the algorithm output is dependent on the initial starting point. To this end, we study the effects of multiple random initial starting points. We compare the variability in clustering output due to starting point versus variability due to noise effects to give a sense of the magnitude of variability. We set a single FACET output as the reference output, and compared its JI similarity with outputs from different starting points and increasing noise effects. The boxplots in Fig. 5.5a, b show the effect of initial starting point versus noise on FACETS. At 0 noise rate, the variability in JI similarity is due to initial starting point. Given the fact that high throughput datasets are inherently noisy (as mentioned above), the variability due to starting point is less significant. In addition, Fig. 5.4a–c show the effect of starting points with respect to gold standard datasets when one observes the similarity at 0 noise rate.

Convergence. Figures 5.6a, b show the functional reassignments after the i-th iteration. We conduct the tests on varying types of datasets with $n = 6$. We also vary the number of facets per atlas ($n = 2$ to 6) on the global human network. All tests converge in less than 9 rounds, demonstrating FACETS' ability to converge quickly to a solution. Larger datasets such as the human network require more iterations to complete. The number of iterations required also tends to increase with the number of facets n.

Statistical significance of FACETS clusters. We utilize the $p - value$ bounds described in Chap. 4 to evaluate the statistical significance of FACETS clusters.

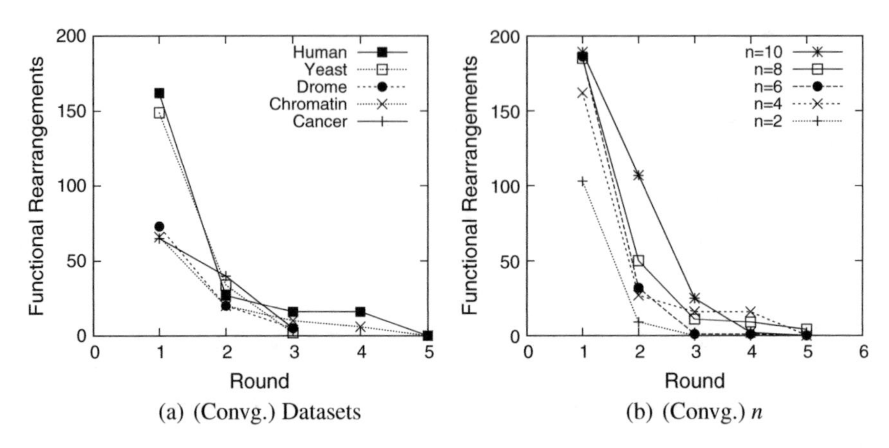

(a) (Convg.) Datasets (b) (Convg.) n

Fig. 5.6 Rate of convergence of FACETS algorithm

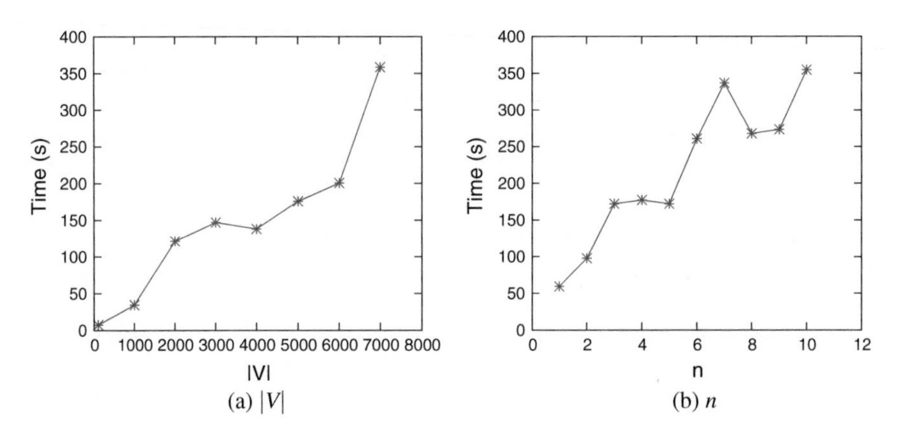

(a) $|V|$ (b) n

Fig. 5.7 Running time of FACETS algorithm

Table 5.3 shows the most significant upper bound $p - value$ scores and the cluster size needed to satisfy the bound. We note that all of the clusters we obtain from FACETS summary are at least as large as the required size needed to satisfy the upper bound. This indicates that FACETS clusters are more siginificant than randomly drawn subgraphs when assessed by their subgraph densities.

Running time. Figures 5.7a, b plot the running times of FACETS with varying network sizes $|V|$ and facet count n. Observe that the running time of FACETS on the largest network (human) is less than 3 min with $n = 11$ and less than a minute with $n = 2$.

Varying parameters of graph clustering methods yield delta differences. We evaluate whether graph clustering approaches can generate functionally orthogonal decompositions by varying their parameters. Figures 5.8a–c report the effect of

Table 5.3 The p-value significance of FACETS clusters

Facet	Cluster size	Maximum size	p-value
0	5	1.76612927	6.49E-07
2	5	1.76612927	6.49E-07
4	5	1.76612927	6.49E-07
5	5	1.76612927	6.49E-07
3	8	1.783769936	7.36E-07
0	4	1.940572357	1.99E-06
1	4	1.940572357	1.99E-06
1	4	1.940572357	1.99E-06
1	4	1.940572357	1.99E-06
1	4	1.940572357	1.99E-06
1	4	1.940572357	1.99E-06
1	4	1.940572357	1.99E-06
2	4	1.940572357	1.99E-06
2	4	1.940572357	1.99E-06
3	4	1.940572357	1.99E-06
3	4	1.940572357	1.99E-06
4	4	1.940572357	1.99E-06
4	4	1.940572357	1.99E-06
5	4	1.940572357	1.99E-06
5	4	1.940572357	1.99E-06
5	4	1.940572357	1.99E-06
5	4	1.940572357	1.99E-06
5	4	1.940572357	1.99E-06
5	4	1.940572357	1.99E-06
1	6	2.037508859	3.44E-06
1	5	2.037508859	3.44E-06
3	6	2.037508859	3.44E-06
3	5	2.037508859	3.44E-06
5	5	2.037508859	3.44E-06
5	5	2.037508859	3.44E-06
5	11	2.152382327	6.25E-06
1	5	2.381391052	1.83E-05

varying parameters on the JI similarity scores between the gold standard decompositions and the clustering output of the human network. Despite varying parameters to generate different decompositions, the decompositions are still largely from a single perspective – those of protein complexes. This is indicated by highest clustering similarity to the `mips` dataset. We suggest that ignoring the clustering of protein complexes, which are dense modules, causes significant drop in the objective function

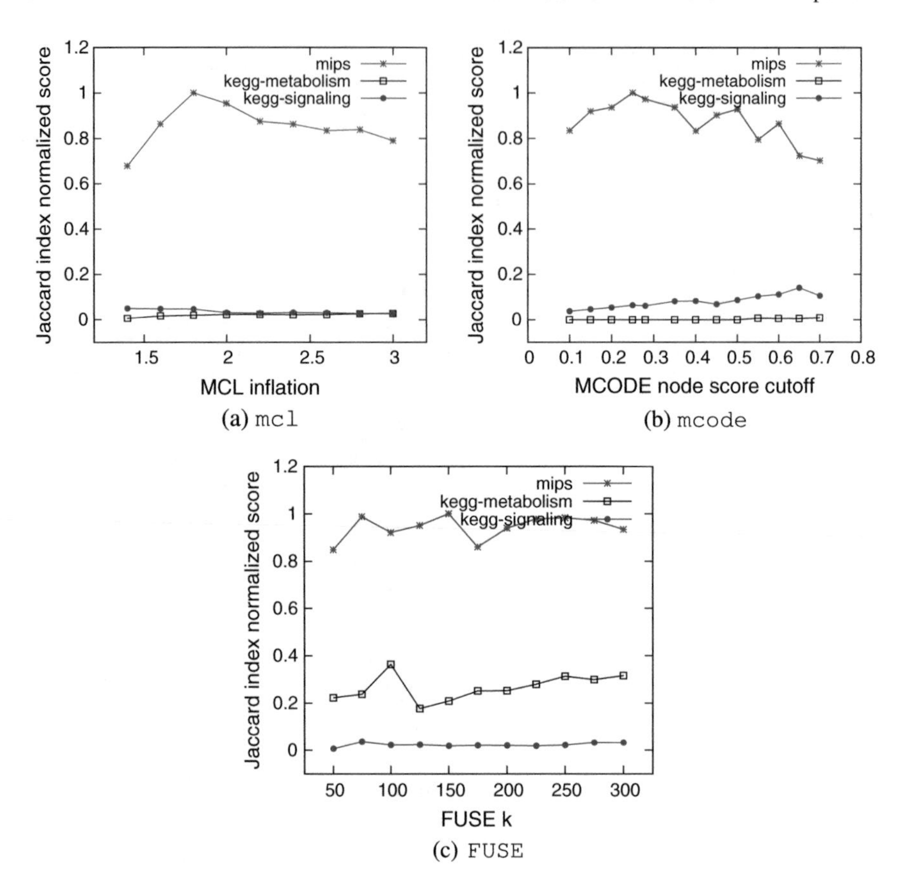

Fig. 5.8 Varying parameters of clustering methods

score of clustering methods. Weaker clusters of other decompositions are hidden by the dominant complexes and are unlikely to be prioritized over complexes by adjusting the clustering parameters.

Comparison with GO DAG. Finally, we evaluate whether FUSE and FACETS generated summaries are superior to a baseline of simply taking GO terms of a certain level in the GO DAG. The baseline of GO terms at Level 2, for instance, represent the GO terms that are located in level 2 of the GO DAG. We assume that by taking GO terms at a particular level and forming clusters using these terms, we could construct a set of clusters that represent a functional summary of the PPI network. To evaluate such baseline against FUSE and FACETS summaries, we consider two measures that evaluate the quality of clusters obtained. First, we use the *average cluster coherence* score to measure the average structural density of a cluster. To this end, it is simply the average of the *ratio association* [22] scores of clusters in a summary. The average

Table 5.4 Comparing GO terms at a particular level in the GO DAG

Method	#Clusters	Distinctiveness	Avg. cluster coherence
GO Terms @ Level 2	34	0.155	0.006
GO Terms @ Level 3	82	0.130	0.012
GO Terms @ Level 4	290	0.119	0.027
GO Terms @ Level 5	534	0.106	0.023
GO Terms @ Level 6	797	0.121	0.029
FACETS	350	0.725	**0.513**
FUSE	150	**0.835**	0.359

cluster coherence score is used to evaluate the modularity of clusters in a summary. Distinctiveness (the inverse of the redundancy metric) measures the lack of cluster overlap in the summary, and is defined as:

$$distinctiveness(\Theta) = \frac{\left| \bigcup_{C(u) \in S_\Theta} V(u) \right|}{\sum_{C(u) \in S_\Theta} |V(u)|} \quad (5.5)$$

where $distinctiveness(\Theta) \in [0, 1]$. A high distinctiveness score implies that few overlap exists between clusters (i.e., more interpretable summary), while a very low distinctivesness score implies that many of the clusters are significantly overlapping.

Table 5.4 presents our findings. Observe that both FUSE and FACETS summaries have significantly higher distinctiveness and average cluster coherence scores compared to the baseline. The average cluster coherence is at least 10 times greater than that obtained from the baseline (clusters of the summaries are strongly connected), while the distinctiveness is almost 5 times greater (less overlap between clusters). In general, summaries generated using our methods form modules or clusters that are much more interpretable and structurally significant.

5.6 Case Study: Human Autophagy System

To illustrate the utility of multi-faceted decomposition, we analyze the functional organization of human autophagy system. Autophagy is the process where proteins and organelles are degraded [23]. Autophagosomal vesicles deliver such components to the lysosome or vacoule, where they are degraded. The autophagy system thus regulates the expression of proteins, as well as removing defective components. Multiple diseases arise from the dysfunction of the autophagy system. It is thus relevant to study the organization of such system. The functional map of this system

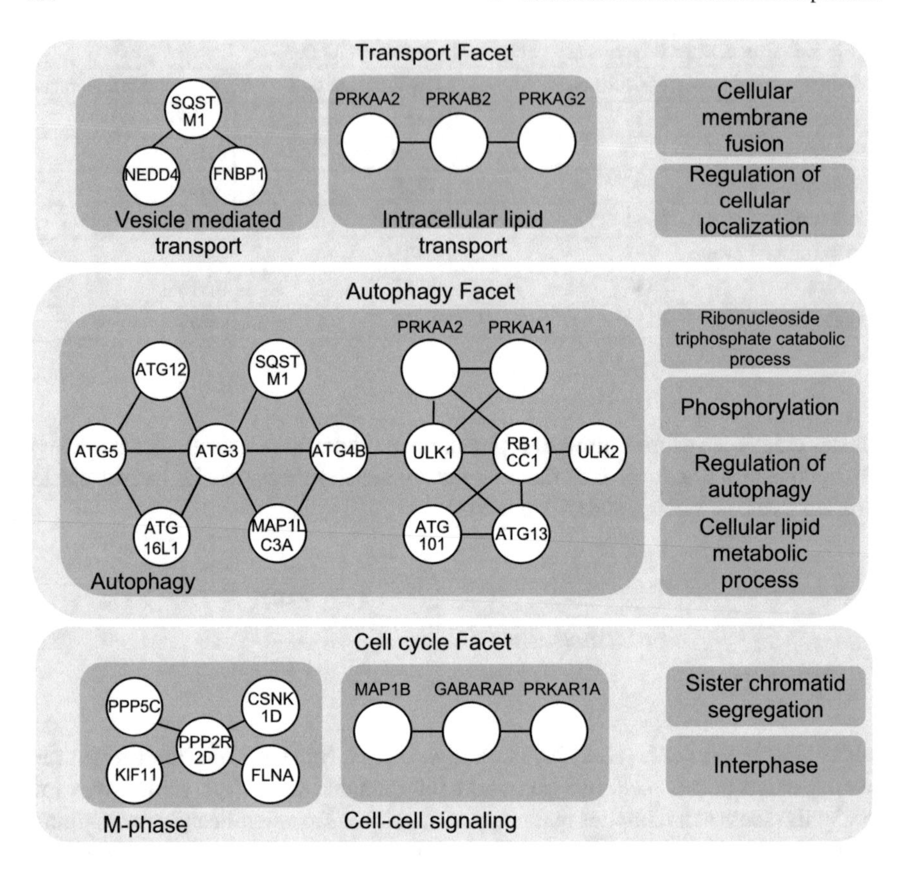

Fig. 5.9 Multiple facets (subset) illustrating the functional organization of the human autophagy network under different perspectives

was manually constructed in [24]. The authors found many genes are significantly implicated in vesicle transport, GTPase signaling, proteolysism ubiquitination and phosphorylation.

We generate the facets of the human autophagy network ($n = 6$), and a subset of the results is shown in Fig. 5.9. The automatically generated facets show the pertinent roles of vesicle transport and lipid membrane metabolism in autophagy, which is consistent with the manually constructed map. For instance, we observe that transport role is a key facet of the autophagy system. This correlates with the finding in [24] that more than half of the proteins in the system are linked to vesicle transport. The genes implicated in vesicle function include `NEDD4`, `SQSTM1` and `FNBP1`. `NEDD4` has been implicated in endosomal protein degradation [25]. The `SQSTM1` has been previously found to be involved in recruitment of ubiquinated cargo [25].

Additionally, the network can also be clustered from the perspective of cell cycle and apoptosis regulation modules, which is not depicted in the manual map. The mTOR inhibition occurs in association with MAP1B. Meanwhile, the GABARAP protein is an ortholog of the key autophagy associated protein ATG8, which is shown to be implicated in cell growth related signaling [24]. Other genes are found to be implicated in cell growth control, including STK3/MST2 and STK4/MST1 – the components of the Hippo kinase complex.

In summary, having multiple perspectives allow explanation of the organization of a network from several angles.

5.7 Conclusions

In this chapter, we describe a data-driven and generic algorithm called FACETS for generating multi-faceted functional decompositions of a PPI network, providing multiple perspectives of the functional organization landscape of the network. Our experimental validation with real-world PPI networks demonstrates effectiveness of FACETS in generating functionally distinctive facets. These distinctive facets have higher relevance to real life datasets compared to single decomposition-based graph clustering techniques discussed in Chap. 3.

References

1. B.-S. Seah, S.S. Bhowmick, C.F. Dewey, FACETS: multi-faceted functional decomposition of protein interaction networks. Bioinformatics (Oxford, England) **28**, 2624–2631 (2012)
2. G.D. Bader, C.W.V. Hogue, An automated method for finding molecular complexes in large protein interaction networks. BMC Bioinform. **4**, 2 (2003)
3. N.J. Krogan, G. Cagney, H. Yu, G. Zhong, X. Guo, A. Ignatchenko, J. Li, S. Pu, N. Datta, A.P. Tikuisis, T. Punna, J.M. Peregrín-Alvarez, M. Shales, X. Zhang, M. Davey, M.D. Robinson, A. Paccanaro, J.E. Bray, A. Sheung, B. Beattie, D.P. Richards, V. Canadien, A. Lalev, F. Mena, P. Wong, A. Starostine, M.M. Canete, J. Vlasblom, S. Wu, C. Orsi, S.R. Collins, S. Chandran, R. Haw, J.J. Rilstone, K. Gandi, N.J. Thompson, G. Musso, P. St Onge, S. Ghanny, M.H.Y. Lam, G. Butland, A.M. Altaf-Ul, S. Kanaya, A. Shilatifard, E. O'Shea, J.S. Weissman, C.J. Ingles, T.R. Hughes, J. Parkinson, M. Gerstein, S.J. Wodak, A. Emili, J.F. Greenblatt, Global landscape of protein complexes in the yeast Saccharomyces cerevisiae. Nature **440**, 637–643 (2006). Mar
4. B.-S. Seah, S.S. Bhowmick, C.F. Dewey, H. Yu, FUSE: a profit maximization approach for functional summarization of biological networks. BMC Bioinform. **13**(Suppl 3), S10 (2012). Jan
5. Z. Qi, I. Davidson, A principled and flexible framework for finding alternative clusterings, in *Proceedings of the 15th ACM SIGKDD international conference on Knowledge discovery and data mining - KDD '09*, (ACM Press, New York, 2009), p. 717
6. D. Niu, J.G. Dy, M. I. Jordan, Multiple non-redundant spectral clustering views, in *Proceeding of the 27th International Conference on Machine Learning - ICML '10, Haifa, Israel* (2010)
7. Y. Cui, X.Z. Fern, J.G. Dy, Non-redundant Multi-view Clustering via Orthogonalization, vol. 3, (IEEE, 2007)
8. C.C. Kiri Wagstaff, Clustering with instance-level constraints, in *Proceedings of the Seventeenth International Conference on Machine Learning* (2000)

9. N.N. Rich Caruana , Mohamed Elhawary, Meta clustering, in *IEEE International Conference on Data Mining* (2006)
10. G. Agarwal, D. Kempe, Modularity-maximizing graph communities via mathematical programming. Eur. Phys. J. B **66**, 409–418 (2008). Nov
11. C. Massen, J. Doye, Identifying communities within energy landscapes. Phys. Rev. E **71**, 046101 (2005). Apr
12. C. Kingsford, S. Navlakha, Exploring biological network dynamics with ensembles of graph partitions, in *Pacific Symposium Biocomputer* (2010), pp. 166–77
13. S. Navlakha, J. White, N. Nagarajan, M. Pop, C. Kingsford, Finding biologically accurate clusterings in hierarchical tree decompositions using the variation of information. J. Comput. Biol. J. Comput. Mol. cell Biol. **17**, 503–516 (2010). Mar
14. A. Jagota, Approximating maximum clique with a Hopfield network. IEEE Trans. Neural Netw. Publ. IEEE Neural Netw. Counc. **6**, 724–735 (1995). Jan
15. Y. Botton, L. Bengio, Convergence properties of the k-means algorithms, in *In Advances in Neural Information Processing Systems 7* (1994)
16. S. Kerrien, Y. Alam-Faruque, B. Aranda, I. Bancarz, A. Bridge, C. Derow, E. Dimmer, M. Feuermann, A. Friedrichsen, R. Huntley, C. Kohler, J. Khadake, C. Leroy, A. Liban, C. Lieftink, L. Montecchi-Palazzi, S. Orchard, J. Risse, K. Robbe, B. Roechert, D. Thorneycroft, Y. Zhang, R. Apweiler, H. Hermjakob, IntAct-open source resource for molecular interaction data. Nucl. Acids Res. **35**, D561–D565 (2007). Jan
17. A. Ben-Hur, A. Elisseeff, I. Guyon, A stability based method for discovering structure in clustered data, in *Biocomputing 2002 - Proceedings of the Pacific Symposium*, (World Scientific Publishing Co. Pte. Ltd., Singapore, 2001), pp. 6–17
18. H.W. Mewes, D. Frishman, U. Güldener, G. Mannhaupt, K. Mayer, M. Mokrejs, B. Morgenstern, M. Münsterkötter, S. Rudd, B. Weil, MIPS: a database for genomes and protein sequences. Nucl. Acids Res. **30**, 31–34 (2002). Jan
19. C.G. Rivera, R. Vakil, J.S. Bader, NeMo: Network Module identification in Cytoscape. BMC Bioinform. **11**(Suppl 1), S61 (2010). Jan
20. E.I. Boyle, S. Weng, J. Gollub, H. Jin, D. Botstein, J.M. Cherry, G. Sherlock, GO::TermFinder–open source software for accessing Gene Ontology information and finding significantly enriched Gene Ontology terms associated with a list of genes. Bioinformatics (Oxford, England) **20**, 3710–3715 (2004)
21. G.T. Hart, A.K. Ramani, E.M. Marcotte, How complete are current yeast and human protein-interaction networks? Gen. Biol. **7**, 120 (2006). Jan
22. P.K. Chan, M.D.F. Schlag, J.Y. Zien, *Spectral K-way Ratio-cut Partitioning and Clustering* (ACM Press, New York, 1993)
23. N. Mizushima, B. Levine, A.M. Cuervo, D.J. Klionsky, Autophagy fights disease through cellular self-digestion. Nature **451**, 1069–1075 (2008). Feb
24. C. Behrends, M.E. Sowa, S.P. Gygi, J.W. Harper, Network organization of the human autophagy system. Nature **466**, 68–76 (2010). July
25. I. Novak, V. Kirkin, D.G. McEwan, J. Zhang, P. Wild, A. Rozenknop, V. Rogov, F. Löhr, D. Popovic, A. Occhipinti, A.S. Reichert, J. Terzic, V. Dötsch, P.A. Ney, I. Dikic, Nix is a selective autophagy receptor for mitochondrial clearance. EMBO Rep. **11**, 45–51 (2010). Jan

Chapter 6
Differential Functional Summarization

In the preceding chapters, we have focused our discussions on clustering and summarizing *static* biological networks. This limits the power to summarize the complex behavior of a biological network as it is dynamic and responds to both environmental and genetic factors. The complexity associated with understanding the dynamic behavior of biological systems is well known, hence this is a challenging problem that requires careful investigation. To this end, in this chapter we present `DiffNet` [1] towards generating *differential* functional summary between two snapshots of a specific type of biological network under contrasting environmental conditions.

6.1 Background

High-throughput mapping of genetic interaction networks of a set of genes is an important and emergent research problem [2]. The networks constructed with these methods, however, only represent a *static* "snapshot" of the genetic interaction map under a particular context or condition. Recent studies have shown that genetic interaction maps are in fact dynamic and context-dependent [3]. Consequently, there is a growing interest in studying the system-wide responses of interaction networks following environmental or condition change [4, 5]. For instance, one may be interested in elucidating the genetic interaction differences between cancer cells and normal cells. Specifically, some interactions may appear or disappear in the disease state, intensity of some interactions may alleviate or aggravate when in disease state compared to healthy condition, and others may remain strong irrespective of the state.

One representative method that has been recently proposed for mapping the genetic interaction responses following environment change is the `dE-MAP` approach [6]. In this method, two static gene interaction networks [2] for each condition are first obtained using the *epistatic miniarray profile* (`E-MAP`) approach [7], which constructs a quantitative genetic interaction landscape of *S. cerevisiae* by first

S.S. Bhowmick and B.-S. Seah, *Summarizing Biological Networks*,
Computational Biology 24, DOI 10.1007/978-3-319-54621-6_6

identifying a set of genes of interest. Double mutant strains of all pairwise genes from this set of genes are then grown and their colony size measured. Genetic interaction occurs between a pair of mutant genes when one observes greater or lesser than expected colony growth rate when compared to their respective single mutant strains. When the growth rate is greater than expected, the interaction is deemed *positive* (alleviating); when it is lesser, it is deemed *negative* (aggravating). Using the two static E-MAP networks, a *differential network* (dE-MAP network) is then computed that maps the interaction differences between the two static networks. For example, in [6], *S. cerevisiae* E-MAP networks are obtained for cells grown under two conditions: (a) cells which are treated with methyl methanesulfonate (MMS), a well known DNA-damaging agent and (b) cells which are untreated. Large-scale genetic interaction network among 418 yeast genes is quantitatively extracted using the E-MAP method under the MMS-treated condition (stressed) and untreated condition (unstressed) and the differential network that maps the genetic interaction changes due to MMS challenge is computed. Figure 6.1 depicts an example of a differential network (partial view) that is obtained from two static E-MAP networks under MMS-treated and untreated condition.

Naturally, it is important to analyze this differential network to investigate the system-wide impact of the DNA-damaging agent on the functional roles of various components. Consequently, the authors obtained physical protein-protein interactions corresponding to these genes and performed graph clustering to find protein

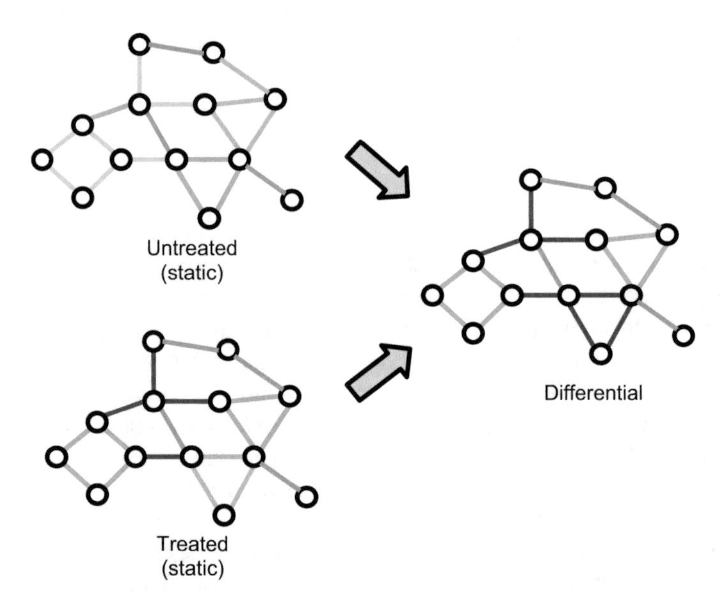

Fig. 6.1 The differential network that arises from two static E-MAP networks under different conditions. *Red* interactions – positive differential; *green* interactions – negative differential

complexes[1] enriched with differential interactions. The functional identity of each cluster is then *manually*[2] determined. Particularly, the authors concluded that these complexes tend to be stable across conditions and differential interactions largely lie between complexes, rather than within complexes. Unfortunately, modules constructed in this manner poorly represent the functional responses of the differential network. Hence, to find a functional response, the authors manually selected a subset of 31 genes associated with `DNA repair` to test for differential interaction enrichment, concluding that `DNA repair` is a pertinent functional response following `MMS`-treatment.

6.2 Motivation and Overview

It is time-consuming, laborious and error-prone to perform large-scale analysis of `dE-MAP` interactomes to map all pertinent functional responses. In this chapter, we present a technique called `DiffNet` [1] that addresses this impediment by automatically constructing a high quality differential summary of two `E-MAP` networks under environmental change. Figure 6.2 highlights some of these functional modules that are differentially effected by the DNA-damaging agent.

At first glance, the aforementioned failure of traditional graph clustering techniques to capture differential summaries in its modules may seem surprising. However, as we shall see in Sect. 6.6, these techniques are largely designed for static networks and are less suitable for differential networks that contain both positive and negative weights. Furthermore, since most methods rely solely on topology of the network, there is also no guarantee that each cluster corresponds well to a representative biological function response. In fact, as remarked earlier, in [6] the functional identity of each cluster following graph clustering is manually determined. Furthermore, the authors failed to assign function to a significant number of these clusters.

In fact, algorithms that perform genome-wide functional analysis of gene responses under multiple conditions have been proposed in the literature [12–14]. Particularly, these approaches perform functional analysis based on the expression levels of genes. In contrast, in our problem we focus on genome-wide functional analysis of the *gene interactions* and their responses.

Given the differential network generated from `dE-MAP` interactions, `DiffNet` greedily constructs a *differential summary* comprising of a set of *skewed* and *coherent functional subgraphs*, representing significant functional responses following environment or condition change. Specifically, it leverages `GO` annotations to identify these functional subgraphs, each of which represents a group of interactions corresponding to a specific biological function. A key characteristic of these functional subgraphs is that the interactions together respond *significantly* in one direction,

[1] The topology of the differential network can be mined to identify gene clusters using techniques such as [8–10].

[2] A function can also be associated with each cluster by leveraging a *functional enrichment* technique [11].

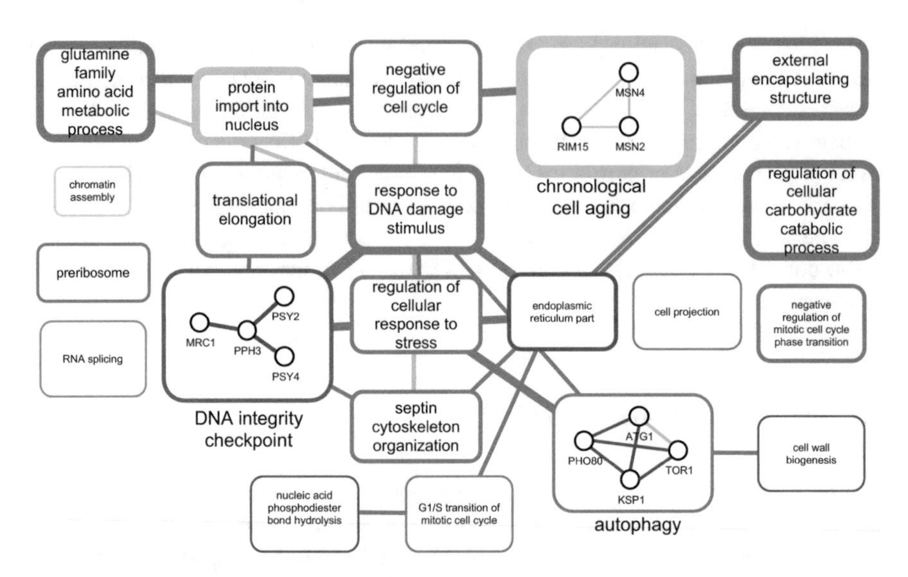

Fig. 6.2 Differential functional summary of MMS-induced/untreated yeast de-MAP network in [6]. The color of the functional modules and gene interactions indicate either positive differential (*red*) or negative differential (*green*). The thickness of the lines indicate the strength of the differential response. Gene interaction subgraphs of selected functional modules are also shown. Edges between functional modules depict differential interactions that occur between functional modules. The thickness of these edges represent the skewness of the differential interactions between a pair of functional modules. The most significant of such edges are shown

either positively or negatively, to the condition change. That is, unlike standard graph clustering methods, DiffNet is specifically designed to handle *differential interactions*, which can be positively or negatively weighted. Figure 6.3 illustrates the idea of the DiffNet algorithm.

6.3 Functional Subgraphs in a Differential Network

In this section, we define the notion of functional subgraphs in a differential network instead of a static network. We begin by describing the process of constructing differential networks.

6.3.1 Constructing Differential Networks

The set of genes of interest together with their genetic interactions can be modeled as a gene-gene interaction network, denoted by $G = (V, E, w)$, where V is a set of genes selected for E-MAP study, E denotes the pairwise interactions between

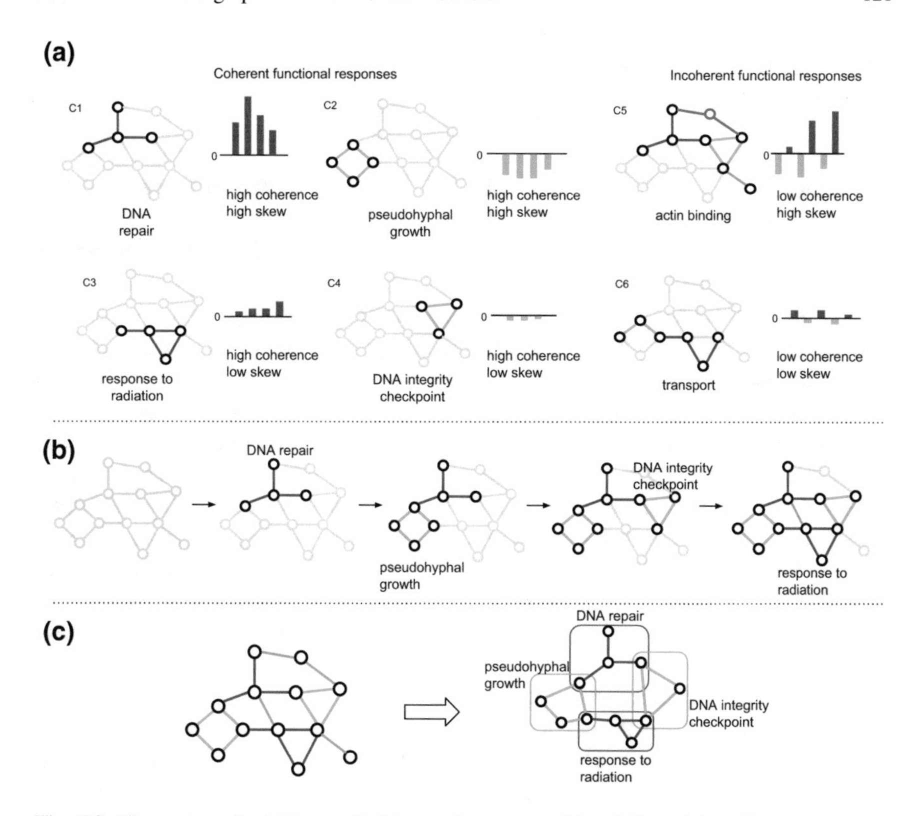

Fig. 6.3 Illustration of `DiffNet`. *Red* interactions are positive differential, while *green* interactions are negative differential. **a** A functional subgraph represents interacting genes that share a specific function (e.g., C1 represents gene interactions associated with `DNA repair`). A coherent functional subgraph has differential interactions that mostly respond in one direction. We say that a functional subgraph has high skew if the differential interaction weights have high magnitude; it has high coherence when the interactions largely respond in one direction. A functional subgraph with high coherence and skew represents a concerted, significant functional response due to the condition change. **b** The `DiffNet` algorithm implements a greedy heuristic that selects, at each iteration, the functional subgraph with highest coherence and skew from the remaining unselected interactions. **c** The output of `DiffNet` is a decomposition that summarizes the relevant functional responses due to condition change

genes, and w is a function that assigns each pairwise interaction $e \in E$ a weight that represents its interaction strength. In `E-MAP` studies, $w(e)$ of $e \in E$ is given by its genetic interaction score $S\text{-}score$ [7]. A positive $S\text{-}score$ indicates the degree of alleviating interaction between the two genes whereas a negative $S\text{-}score$ indicates the degree of aggravating interaction. Therefore, $w(e)$ can be positive or negative.

Consider now two `E-MAP` networks $G_t = (V, E, w_t)$ and $G_c = (V, E, w_c)$ that represent two conditions: (a) the treated condition (G_t) and (b) the untreated condition (G_c). Observe that G_t and G_c share the same set of vertices and pairwise interactions.

Given G_t and G_c, the differential network of G_t and G_c is a graph $G_d = (V, E, w_d)$ such that $\forall e \in E$:

$$w_d(e) = \left(1 + e^{-\frac{w_t(e)-w_c(e)}{|w_c(e)|}}\right)^{-1} - 0.5 \tag{6.1}$$

We apply the logistic function $(1 + e^{-x})^{-1}$ (shifted by 0.5 to make it an odd function) to "clip" potentially large magnitudes of differential responses. This is inspired by a similar approach used in activation functions in neural networks to bound the response of signals [15].

Intuitively, a differential network models gene interaction responses due to condition change. The differential weight $w_d(e)$ represents the normalized difference in $S\text{-}scores$ between the two conditions for a pair of genes represented by e. We call $w_d(e)$ *positive differential* when $w_d(e) > 0$, and *negative differential* when $w_d(e) < 0$. A positive (*resp.* negative) differential response indicates increased alleviating (*resp.* aggravating) interaction between the two genes in treated condition compared to untreated condition. The magnitude of $w_d(e)$ reflects the strength of interaction response due to condition change. Figure 6.4 shows a toy differential network of positive (red) and negative (green) differential interactions. Grey colored interactions do not respond to condition change (i.e., $w_d(e) \approx 0$). The interaction between `RAD52` and `SIN3`, for instance, has a positive differential response due to condition change.

It is worth noting that the above definition of differential interaction w.r.t DNA damage-induced `dE-MAP` network is consistent with the one in [6]. Specifically, a positive differential interaction indicate DNA damage-induced lethality, while a negative differential interaction indicate inducible epistasis or suppression. Importantly, the differential response does not distinguish, for example, one that goes from negative to positive from one that goes from positive to more positive. Although the former is arguably more interesting, the latter still is biologically significant because it indicates a significant response due to treatment.

Although we now have a model of individual gene-gene interaction responses due to condition change, it remains unclear how one *automatically* infers broader, systemic functional responses from these detailed interactions. This issue is pertinent

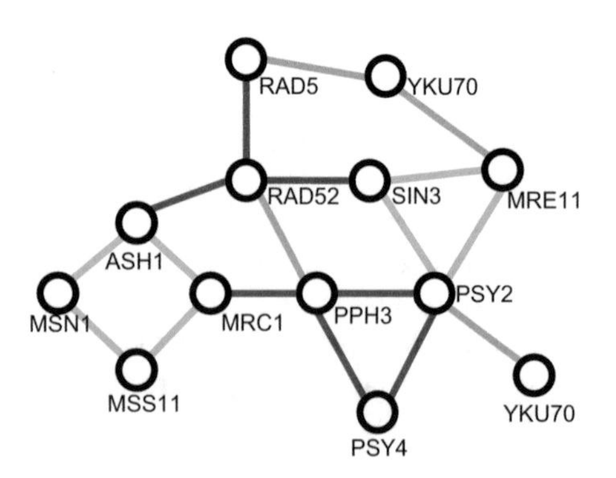

Fig. 6.4 A toy differential network of gene interactions

in high-throughput experiments, which often generate thousands, even millions, of interacting genes within a single experiment. Hence we present our approach to model responses due to condition change from a functional perspective.

6.3.2 Functional Subgraphs

We begin by modeling a systemic functional response by a subgraph of functionally-similar gene interactions (i.e., a set of genes of a specific function and their interactions). Let $\Delta = \{T_1, T_2, \ldots\}$ be a set of GO terms in the Gene Ontology. This represents the set of biological functions relevant to our study. Every gene $v \in V$ is annotated with zero or more biological functions in Δ. Then a *functional subgraph*, denoted by $C_T = (V_T, E_T)$, is a subgraph of G_d such that: (a) C_T is a subgraph of G_d induced by V_T, and (b) every gene $v \in V_T$ shares a function $T \in \Delta$. For instance, the subgraph $C1$ in Fig. 6.3a is a functional subgraph of genes sharing the `DNA repair` function. One can see that a functional subgraph models the interaction responses of genes with a specific function as a whole.

We evaluate each functional subgraph C_T with the *skewness* and *coherence* measures. We say that a functional subgraph is *skewed* if its interactions significantly respond to condition change (i.e., the interactions in the subgraph are significantly positive or negative differential). Analogous to individual gene interactions, we call a subgraph $C_T = (V_T, E_T)$ *positively skewed* if the sum of its edge weights, defined as $skew(C_T) = \sum_{e \in E_T} w_d(e)$, is greater than 0; it is *negatively skewed* if the sum of its edge weights is less than 0, i.e., $skew(C_T) < 0$. The greater the value of $skew(C_T)$, the more the interactions of C_T respond to condition change.

We say that a functional subgraph is *coherent* if its interactions are largely skewed in one direction (either positive or negative differential). Figure 6.3a depicts the coherence of subgraphs of the toy network in Fig. 6.4. Consider the subgraph representing `DNA repair` function. It is coherent because it consists of interactions that are skewed towards positive differential in tandem. Intuitively, this would mean that the `DNA repair` function, as a whole, has increased alleviating response due to the condition change. Meanwhile, the subgraph representing `transport` has a mix of positive and negative differential interactions. There is no clear indication whether the `transport` function is positively or negatively affected by the condition change. We now formally define the notion of *subgraph coherence*. Given a subgraph C_T, $coherence(C_T) \in [0, 1]$ is given by:

$$coherence(C_T) = \frac{max(|\{e : w_d(e) > 0\}|, |\{e : w_d(e) < 0\}|)}{|E_T|} \tag{6.2}$$

The greater the value of $coherence(C_T)$, the more coherent is the subgraph. If $coherence(C_T) = 1$ then it indicates that all interactions are exclusively positive differential or exclusively negative differential.

Figure 6.3a depicts the skewness and coherence of several functional subgraphs. Each bar graph associated with a functional subgraph depicts the differential weight w_d values of the interactions in the subgraph. A high coherence and high skew subgraph has interactions with large w_d values in one direction. On the other hand, a low coherence and low skew subgraph has low w_d values in diverging directions. Consider the following two functional subgraphs: the subgraph of genes sharing the `DNA repair` function (`RAD5, RAD52, SIN3, ASH1`), and subgraph of genes sharing the `transport` function (`MSN1, ASH1, MRC1`, `PPH3`, `PSY4`, `PSY2`). Observe that interactions in the former are positive differential and skewed in one coherent direction, while the latter is not. We are more interested in the former type of subgraphs because it represents a concerted and significant functional response due to the condition change. Generally, functional subgraphs that are high skew and high coherence are informative and represent significant functional responses due to condition change. On the other hand, a subgraph with both low coherence and skew represent function that remain relatively unchanged.

From a statistical point of view, a module constructed from interactions that are unaffected by condition change will have similar interaction distributions, resulting in a coherence score centered around 0 (zero coherence). A high coherence module represents a module with significant change in interaction distribution profile, thus representing a statistically significant module. Biologically, analogous to functional enrichment in gene lists, the statistical significance of high entropy modules means that the function associated with such module exhibit statistically significant interaction response patterns compared to a random function.

Based on the above observation, if one can decompose G_d into a set of highly coherent and skewed functional subgraphs, denoted by $\mathscr{S} = \{C_{T^1}, C_{T^2}, \ldots C_{T^k}\}$, then one can meaningfully obtain a summary representing positive and negative functional responses of G_d due to condition change. We shall later describe how one quantifies the decomposition of G_d based on the coherence and skewness of its functional subgraphs. Consider the decomposition depicted in Fig. 6.3b, c. The network of differential interactions is summarized into a set of functional subgraphs representing the following functional responses – `DNA repair` (positive), `response to radiation` (positive), `DNA integrity checkpoint` (negative) and `pseudohyphal growth` (negative). Each subgraph is coherent and skewed towards either positive or negative differential response.

At this point, it remains unclear how to optimally decompose G_d into a set of coherent and skewed functional subgraphs. To contrast with the previous example, suppose we decompose G_d into $\mathscr{S} = \{$`transport (MSN1, ASH1, MRC1, PPH3, PSY4, PSY2)`, `response to radiation (MRC1, PPH3, PSY4, PSY2)`$\}$. This decomposition poorly summarizes the network in Fig. 6.4 because a significant portion of differential interactions are not captured by the subgraphs in $\mathscr{S}$. The `transport` subgraph also has low coherence.

6.4 Differential Summarization Problem

Given the existence of potentially many possible decompositions of G_d, the problem of *differential summarization* is to identify the *best* decomposition that represents the functional responses in G_d. Suppose we have a set containing all possible functional subgraphs of G_d. Let us denote this set by the universe $\mathscr{E}$. Clearly, some subgraphs will represent meaningful functional responses, while others will be unaffected by the condition change. One would like to choose a subset of $\mathscr{E}$ representing functional responses in G_d that are significantly affected by the condition change. To do this, we must first identify *summarization objectives* that assess the quality of a decomposition of G_d. We argue that a good decomposition of G_d should have the following desirable summary objectives:

- **Subgraph Coherence and Skewness**. A decomposition $\mathscr{S}$ should comprise of functional subgraphs that are significantly coherent and skewed. Recall that our goal is to identify functional regions that significantly respond, either positively or negatively, to condition change. This directly correlates to having coherent and skewed functional subgraphs, and finding $\mathscr{S}$ that maximizes coherence and skewness of its functional subgraphs is desirable. The *differential score* of C_T combines the skewness and coherence of the subgraph as follows:

$$differential(C_T) = coherence(C_T)^{\alpha} \times skew(C_T) \tag{6.3}$$

  where $\alpha \geq 0$ is a parameter controlling the influence of coherence on the differential score. Note that $0 \leq coherence(C_T)^{\alpha} \leq 1$.
- **Edge Coverage**. A good decomposition of G_d should convey key information regarding functional regions affected by condition change. It is natural to prefer a decomposition that covers as much differential interactions in G_d as possible. We introduce the *edge coverage* measure that reflects how well $\mathscr{S}$ represents the differential interactions of G_d. Formally, the *edge coverage* of $\mathscr{S}$ can be expressed as:

$$coverage(\mathscr{S}) = \frac{\left|\bigcup_{C_i \in \mathscr{S}} E_i\right|}{|E|} \tag{6.4}$$

  Intuitively, it indicates the percentage of interactions in G_d that is represented by the subgraphs in $\mathscr{S}$. The wider the coverage, the more representative is the decomposition of the interactions in G_d.
- **Distinctiveness**. Intuitively, two functional subgraphs having disjoint differential interactions are more informative than two redundant subgraphs with identical interactions. Thus, one prefers a decomposition which cleanly partitions G_d into distinctive sets of interactions. We quantify this objective with the *distinctiveness* measure. It quantifies redundancy of functional subgraphs, such that the greater the redundancy, the lower the distinctiveness value. Hence, distinctiveness of $\mathscr{S}$ is 1 if its subgraphs are mutually disjoint. Formally, it is defined as:

$$distinctiveness(\mathscr{S}) = \frac{\left|\bigcup_{C_i \in \mathscr{S}} E_i\right|}{\sum_{C_i \in \mathscr{S}} |E_i|} \tag{6.5}$$

We introduce an optimization model that selects functional subgraphs to maximally cover the set of differential interactions of G_d to maximize the above objective scores. Because the set of possible functional subgraphs can be large, a naive ranking approach of selecting the most significantly coherent and skewed subgraphs can be suboptimal. There is no control on coverage and distinctiveness, leading to significant redundancy in the results. Thus, we propose an optimization model to construct a summary that satisfies all three desirable objective scores. This optimization model can be posed as a *weighted k-set cover* problem [16] of choosing a subset $\mathscr{S} \subseteq \mathscr{E}$ and a set of *remainder subgraphs* $\mathscr{R}$ with cardinality constraint k that minimizes the reciprocal of $differential(\mathscr{S})$. A remainder subgraph $R = (V_R, E_R) \in \mathscr{R}$ is a subgraph of G that is not part of the summary (i.e., $R \cap C_T = \emptyset$ for all $C_T \in \mathscr{S}$). We shall later introduce a penalty for having remainder subgraphs.

Definition 6.1 (*Differential summarization problem*) Let G_d be the differential network of two gene interaction networks, G_c and G_t, under different conditions. Let $U = \bigcup_{C_T \in \mathscr{E}} E_T$ be the universe of differential interactions in G_d where $\mathscr{E}$ is a set of all possible functional subgraphs C_T. The **differential summarization problem** is to identify the *differential decomposition* $\mathscr{S}$ of functional subgraphs and $\mathscr{R}$ of *remainder subgraphs* (representing unselected interactions) by solving the following optimization problem:

$$\begin{aligned}
\arg\min_{\mathscr{S} \cup \mathscr{R}} f(\mathscr{S} \cup \mathscr{R}) = \ & \arg\min_{\mathscr{S}} \sum_{C_T \in \mathscr{S}} differential^{-1}(C_T) + \sum_{R \in \mathscr{R}} r(R) \\
\text{subject to} \quad E \ &= \bigcup_{C_T \in \mathscr{S}} E_T \cup \bigcup_{R \in \mathscr{R}} E_R \\
|\mathscr{S}| + |\mathscr{R}| \ &\leq k
\end{aligned}$$

where the $differential^{-1}(C_T)$ – the reciprocal of the coherence and skewness of C_T – is the cost associated with each functional subgraph $C_T \in \mathscr{S}$, and $r(R) = (|E_R| + 1) \max_{C_T \in \mathscr{E}} differential^{-1}(C_T)$ captures the penalty for not covering the edges of the network.

It can be proven that there is at most one remainder subgraph that can be selected, which is disjoint from all functional subgraphs in $\mathscr{S}$.

Theorem 6.1 *Suppose $\mathscr{S}_0 \cup \mathscr{R}_0$ is an optimal solution. Then $|\mathscr{R}_0| \leq 1$.*

Proof We begin by assuming the contradiction that $|\mathscr{R}_0| > 1$. $\mathscr{R}_0$ covers $\bigcup_{R \in \mathscr{R}_0} V_R$ with cost $|\mathscr{R}_0| \max_{C_T \in \mathscr{E}} differential^{-1}(C_T)$ + $(\max_{C_T \in \mathscr{E}} differential^{-1}(C_T))$ $\sum_{R \in \mathscr{R}_0} |V_R|$. We can replace $\mathscr{R}_0$ with a single remainder subgraph with a lower cost. Let $\mathscr{R}' = \{\bigcup_{R \in \mathscr{R}_0} V_R\}$. The single remainder subgraph $\mathscr{R}'$ covers the same set of vertices with lower cost and set cover cardinality.

Algorithm 4 `DiffNet`.

Input: $G_t = (V, E, w_t)$, $G_c = (V, E, w_c)$, Δ, k
Output: $\mathscr{S}$
1: Let $p_{\max} = 0$
2: **for** $e \in E$ **do**
3: $\quad w_d(e) = (1 + e^{-\frac{w_t(e) - w_c(e)}{|w_c(e)|}}) - 0.5$
4: **end for**
5: Let $G_d = (V, E, w_d)$
6: Let $\mathscr{E} = \emptyset$
7: **for** $T \in \Delta$ **do**
8: $\quad \mathscr{E} \leftarrow \mathscr{E} \cup \{C_T\}$
9: **end for**
10: Let $\mathscr{S} = \emptyset$
11: **repeat**
12: $\quad mincost \leftarrow \infty$
13: $\quad best \leftarrow \emptyset$
14: $\quad$ **for all** $C_T = (V_T, E_T) \in \mathscr{E} \setminus \mathscr{S}$ **do**
15: $\quad\quad SelectedEdges \leftarrow \bigcup_{C \in \mathscr{S}} E$
16: $\quad\quad n \leftarrow |E_T \setminus SelectedEdges|$
17: $\quad\quad f \leftarrow differential^{-1}(C_T)/n$
18: $\quad\quad$ **if** $f < mincost$ and $n > 0$ **then**
19: $\quad\quad\quad mincost \leftarrow f$
20: $\quad\quad\quad best \leftarrow \{C_T\}$
21: $\quad\quad$ **end if**
22: $\quad$ **end for**
23: $\quad \mathscr{S} \leftarrow \mathscr{S} \cup best$
24: **until** $|\mathscr{S}| > k$
25: **return** $\mathscr{S}$

Theorem 6.2 *Suppose* $\mathscr{S}_0 \cup \mathscr{R}_0$ *is an optimal solution. It holds that* $\bigcup_{C_T \in \mathscr{S}_0} V_T \cap \bigcup_{R \in \mathscr{R}_0} V_R = \emptyset$.

Proof Assume by contradiction that $\bigcup_{C_T \in \mathscr{S}_0} V_T \cap \bigcup_{R \in \mathscr{R}_0} V_R \neq \emptyset$. Let $\mathscr{R}' = \{\bigcup_{R \in \mathscr{R}_0} V_R \setminus \bigcup_{C_T \in \mathscr{S}_0} V_T\}$. $\mathscr{S}_0 \cup \mathscr{R}'$ covers the same set of vertices with lower cost.

Note that because of $r(R)$, the aforementioned formulation penalizes a summary that provides low interaction coverage. Also, observe that in principle the above cost function penalizes functional subgraphs with low coherence or skewness scores. The decomposition $\mathscr{S}$ summarizes the key functional responses representing the differences between G_c and G_t. The cardinality constraint k controls the distinctiveness and coverage of the decomposition.

6.5 The `DiffNet` Algorithm

Unfortunately, the differential summarization problem defined in the preceding section is NP-hard because it is posed as a weighted k-set cover problem. Furthermore, it cannot simply be solved by clustering *positive* and *negative* networks

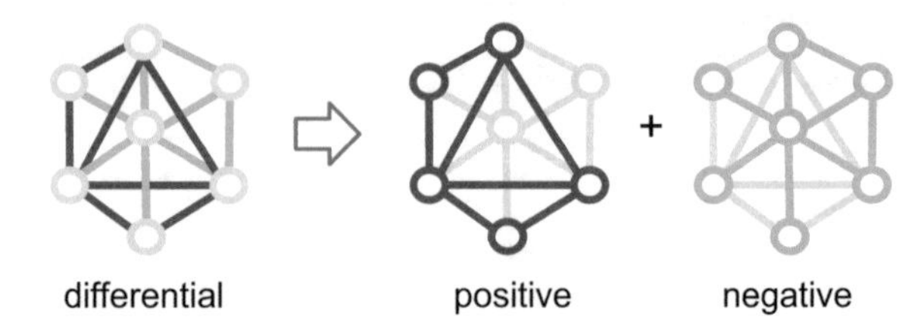

Fig. 6.5 Independently clustering positive and negative edges of differential network

independently. Figure 6.5 shows the separation of a toy differential network into a *positive network* containing only positive differential edges and a *negative network* containing only negative differential edges. When each of the positive or negative network is clustered independently, the information about the other is lost. Consequently, both networks independently show enriched positive and negative interactions. When these interactions are put together, however, the differential interactions of these genes have weak skewness and coherence due to the mixing of positive and negative interactions.

Hence, we describe an algorithm called `DiffNet` that solves this problem heuristically. Here, we adopt a greedy algorithm that admits a H_k-approximation algorithm for the weighted minimum k-set cover problem [16], where $H_k = \sum_{i=1}^{k} \frac{1}{i}$. First, the differential network G_d is computed. Following that, `DiffNet` finds the universe of candidate functional subgraphs of G_d. The basic principle of `DiffNet` is to select, at each iteration, the functional subgraph that gives the best differential score contribution to the existing $\mathscr{S}$. At each iteration, we choose a functional subgraph that maximizes the total differential score. To achieve this, the algorithm maintains a map of interactions of G_d that is represented by currently selected functional subgraphs. For every candidate functional subgraph evaluated for selection, we evaluate its contribution to the remaining unselected interactions. The greedy algorithm then chooses the candidate subgraph that adds the highest differential score to the current summary. This process is iterated until k subgraphs have been selected. Because the penalty of choosing a remainder subgraph is always higher than any functional subgraphs, we let the remainder subgraph, if any, be the last subgraph. Algorithm 4 outlines the pseudocode of the above procedure. Given k passes and the worst case of evaluating $|E|$ edges per subgraph, the proposed algorithm has a worst case complexity of $O(k|\Delta||E|)$.

6.6 Experimental Study

The `DiffNet` algorithm is implemented in Scala. We now present experimental results of the performance of `DiffNet`. The experiments were conducted on a 1.66 GHz Intel Core 2 Duo T5450 machine with 3 GB memory. Unless specified otherwise, we set $k = 45$ and $\alpha = 5.0$.

6.6.1 Functional Analysis of MMS-treated/untreated dE-MAP Network

Using the two E-MAP networks in [6], we construct the differential functional summary associated with MMS treated/untreated genetic interactions. Figure 6.2 shows the differential functional summary of the yeast genetic interactome. We observe significant positive differential functional subgraphs associated with DNA damage and DNA integrity checkpoint. The chronological cell aging genes responsible for stress-resistance – MSN2, MSN4, RIM15 [17] – also undergo significant genetic interaction remodeling following DNA damage. This important and top-scoring functional response is not identified using manual analysis in [6]. The reason why this module could not be detected in [6] is due to their approach of performing cluster analysis on PPI network rather than the differential interaction network itself. Thus, the set of genes, which has less PPI interaction density compared to protein complexes in the PPI network, was missed via conventional cluster analysis. Another type of functional modules that demonstrate significant differential following MMS treatment are pathways related to apoptosis and cell cycle, such as the G1 phase of mitotic cell cycle and cell aging modules. More interestingly, we observe significant negative differential responses in cell projection and cell wall biogenesis functions. The manual functional enrichment study conducted in [6] did not uncover the negative shift of these less obvious groups of genes. The autophagy module, which is a cellular catabolic process, is also seen to be positively activated [18]. Recently, DNA damage has been shown to induce autophagy [18], although the mechanism that triggers remains unclear. Apart from activating autophagy processes, DNA damage is also found to induce actin and septin rearrangement [19]. This is discovered by the differential functional summary, which finds positive activation of septin cytoskeleton organization module.

To contrast the differential functional summary, we also construct a summary of functional subgraphs that shows subgraphs of genes whose genetic interactions remain largely unaltered after MMS treatment. To this end, instead of constructing the differential network G_d, we constructed an "inverse" differential network $G_s = (V, E, w_s)$, such that $w_s = min((w_d(e))^{-1}, \epsilon^{-1})$ where $e \in E$ and ϵ represents a pseudocount that prevents $w_s(e) \rightarrow \infty$. Observe that w_s represents the inverse of normalized $S\text{-}score$ differences. We applied DiffNet on G_s to obtain a landscape of "stable" functional subgraphs, that is, functional subgraphs that are neither strongly positive differential nor strongly negative differential.

Figure 6.6 shows the functional summary of G_s following MMS treatment. The modules represented in this summary could be "housekeeping" processes and modules whose genetic interaction strength remain unaltered regardless of the DNA-damage challenge [6]. For instance, the composition and interaction of the subunits of the RNA polymerase enzyme, a critical module of the cell regardless of cellular context, is unlikely to change. Thus, their genetic interactions should also remain stable. One can make the same argument for preribosome.

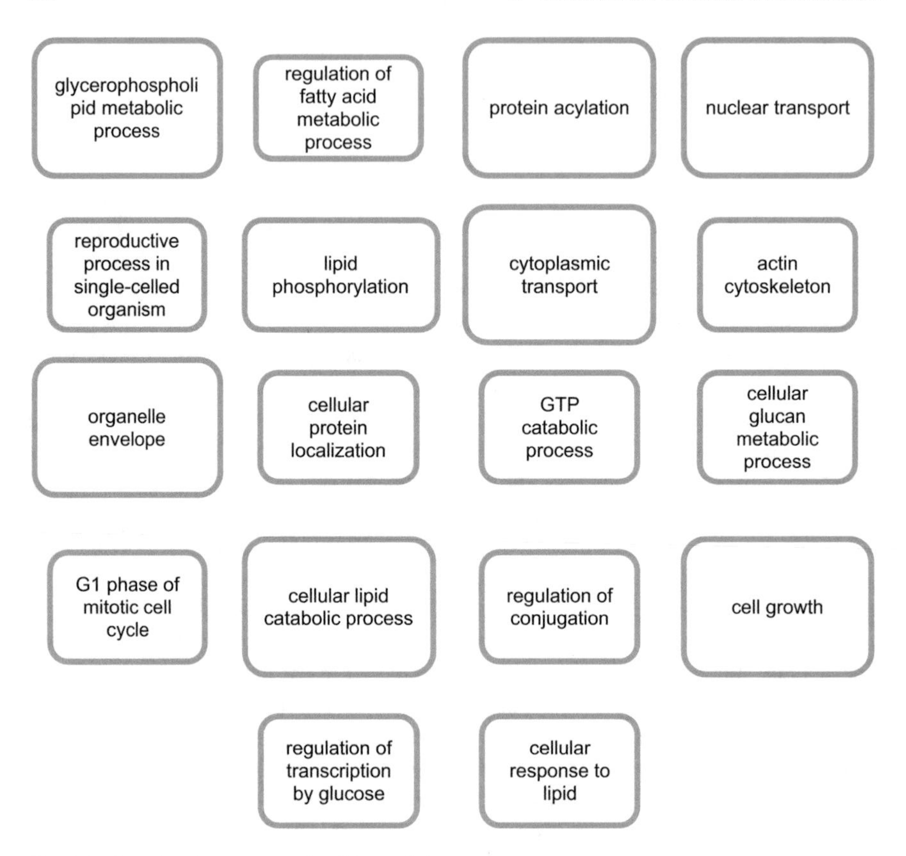

Fig. 6.6 Functional summary of stable modules

6.6.2 *Comparison with Graph Clustering Algorithms*

Since there is no existing technique that automatically generates differential functional summaries, we are confined to compare `DiffNet` with several representative graph clustering methods such as `MCL` [20], Affinity Propagation (`AP`) [9], and `ClusterONE` [10]. We use the dataset in [6] containing 418 genes (393 with annotations). In particular, we chose the `MCL` and `ClusterONE` approaches as a recent evaluation demonstrated that both these methods outperform other graph clustering algorithms on biological networks [10]. Because `MCL` and `ClusterONE` do not accept negative edge weights, they cannot be directly applied to differential networks. To this end, we construct two separate networks from a differential network – (a) a *positive network* containing only positive differential edges and (b) a *negative network* containing only negative differential edges. We assess whether individually clustering both networks using general graph clustering methods, and then aggregating the clusters into one list, could provide results similar to those generated by

DiffNet. For all approaches, we discarded clusters with fewer than 3 genes and selected the 25 best scoring clusters for cluster quality evaluation.

To quantitatively evaluate the quality of the clusters, we introduce several evaluation measures. Given a set of cluster subgraphs $\mathscr{S}$, the *average coherence* and *average skewness* are given by:

$$AvgCoherence(\mathscr{S}) = \frac{1}{|\mathscr{S}|} \sum_{C_T \in \mathscr{S}} coherence(C_T) \tag{6.6}$$

$$AvgSkewness(\mathscr{S}) = \frac{1}{|\mathscr{S}|} \sum_{C_T \in \mathscr{S}} skew(C_T) \tag{6.7}$$

To assess the functional relevance of each cluster, we use the annotation over-representation analysis of the clusters [21]. To this end, the *functional homogeneity* of $\mathscr{S}$ is given by:

$$FuncHomo(\mathscr{S}) = \frac{1}{|\mathscr{S}|} \sum_{C_T=(E_T,V_T) \in \mathscr{S}} -log(p - value(V_T)) \tag{6.8}$$

where $p\text{-}value(V_T)$ is the most significant GO term enrichment $p\text{-}value$ score of the genes in V_T.

Figure 6.7 plots the results of different approaches. Observe that DiffNet is superior to the clustering techniques in the following ways. First, each subgraph in DiffNet has a direct association with a biological function. Recall that functional subgraphs have the constraint that every gene in a subgraph must share a specific function. With graph clustering algorithms such as MCL, each subgraph cluster may contain genes with diverging functions. In that case, it is unclear what biological function the cluster represents. This is quantified by the superior functional homogeneity score of DiffNet. Second, subgraphs in DiffNet have superior coherence compared to other methods. Traditional graph clustering methods are not designed to identify clusters of positive differential interactions and negative interactions. These methods must cluster negative and positive edges independently, and the information encoded in the mixture of positive and negative weights is lost. Third, our method is the second best performer for skewness score. This shows that despite fulfilling multiple summarization constraints, the clusters obtained have high skewness (i.e., high edge weights) scores comparable to general graph clustering methods. Fourth, the 'node-based' decomposition in MCL does not admit overlapping genes. Consider for instance the subgraph $C3$ in Fig. 6.3. If this subgraph is chosen as a cluster in MCL, then the subgraph $C4$ cannot be another cluster because of gene overlap. The 'edge-based' decomposition of DiffNet, which we argue is a more natural way of grouping interaction responses, does not suffer from this problem.

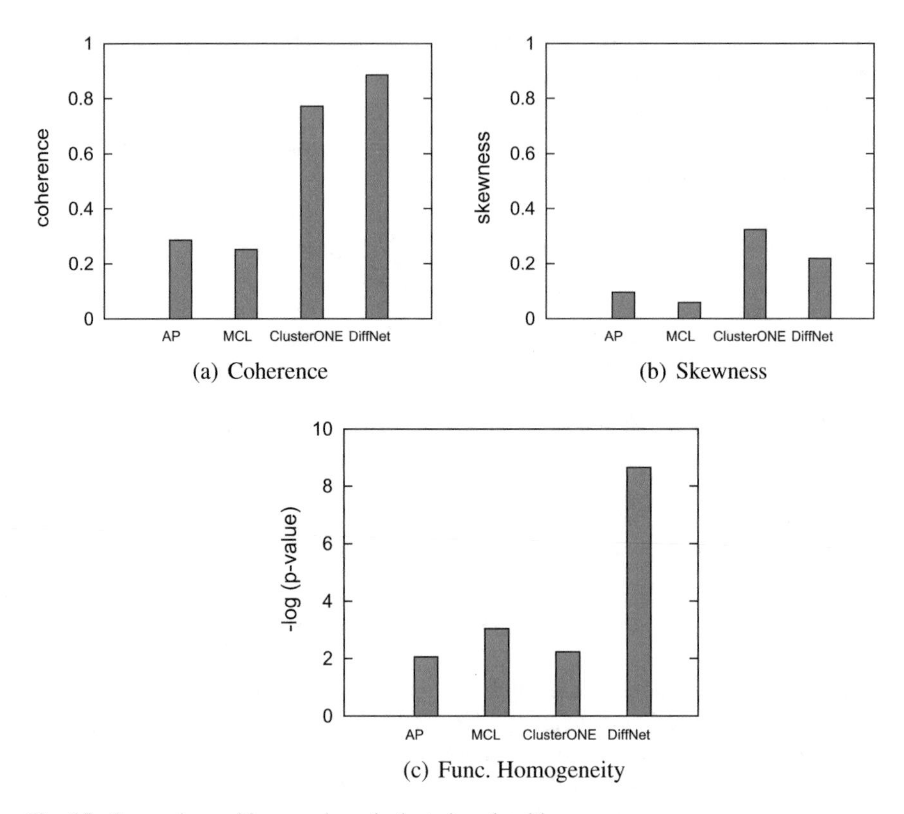

Fig. 6.7 Comparison with general graph clustering algorithms

6.6.3 *Effect of Various Parameters*

Parameter k. Figures 6.8a–d show the effect of k on summary coherence, skewness, coverage and distinctiveness. We observe that k controls the trade-off of summary coverage versus distinctiveness. The higher the value of k, the greater the coverage of functional subgraphs in the summary. However, the increase in coverage reduces the quality of the clusters (lower skewness, coherence and distinctiveness) due to the fact that one must now include lower quality clusters to satisfy the coverage requirement. Note that it is unrealistic to expect the majority of differential interactions to respond significantly to condition change. Thus, full coverage of all interaction responses, especially those that respond weakly, is typically not required in a differential summary.

Parameter α. Figures 6.9a–d show the effect of α on summary coherence, skewness, coverage and distinctiveness. We observe that α directly controls the influence of summary coherence. The higher the value of α, the greater the coherence of functional subgraphs in the summary. The increased coherence, however, comes at a cost. Coverage of the summary is reduced with greater α. This is because the increase

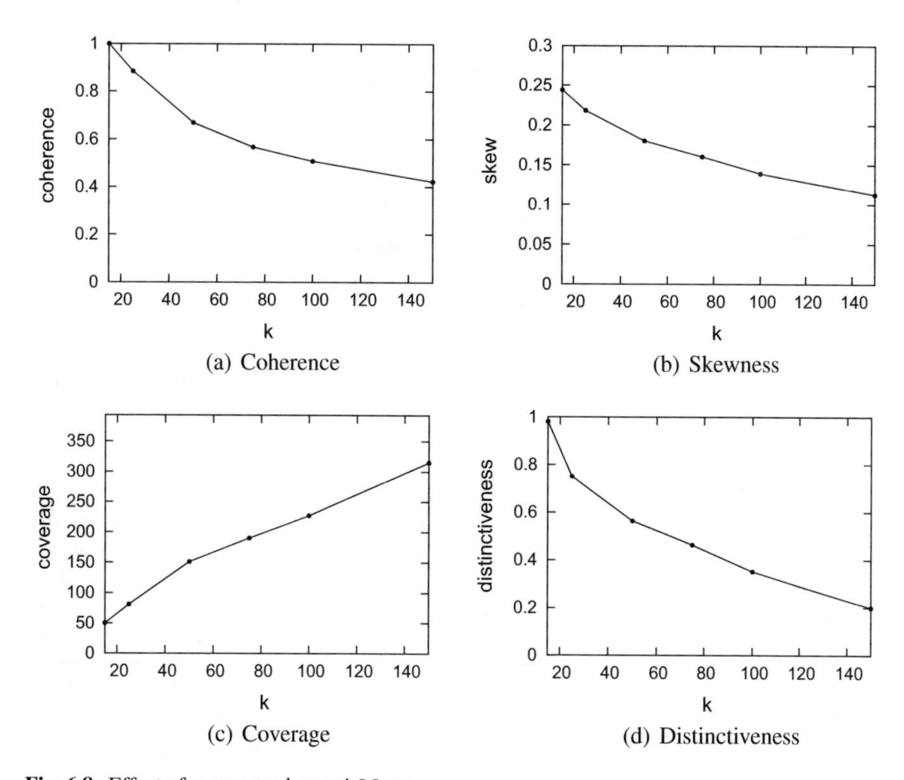

(a) Coherence (b) Skewness

(c) Coverage (d) Distinctiveness

Fig. 6.8 Effect of parameter k on `DiffNet`

penalty for choosing incoherent functional subgraphs reduces the exploration space during decomposition selection. Distinctiveness is also slightly increased with greater α.

Effect of MCL inflation parameter. Figures 6.10a, b show the effect of MCL inflation parameter on summary coherence and skewness. Here we use the recommended range of 1.4–5.0. We observe that the coherence and skewness of functional subgraphs in the summary are stable with varying inflation values. There is, however, a slight increase in coherence and a slight drop in skewness at higher inflation values.

6.6.4 Running Times

We generate synthetic networks by randomly adding nodes and edges to the [6] dataset network until the desired size is obtained. Figures 6.11a, b plot the running times of `DiffNet` of varying network sizes (viewed by number of nodes and edges, respectively). We observe that `DiffNet` scales almost linearly with the number of nodes and edges in the network. A differential network of 2500 nodes is

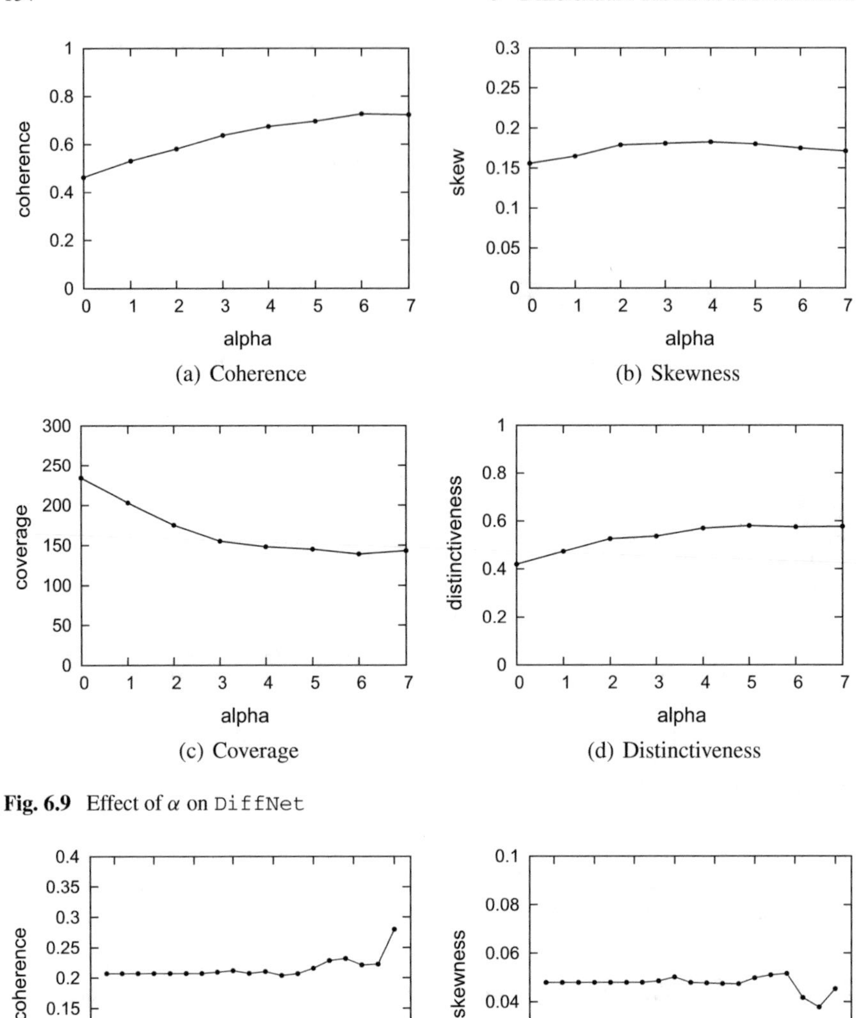

Fig. 6.9 Effect of α on `DiffNet`

Fig. 6.10 Effect of MCL inflation parameter

summarized in less than 3 min. This shows that `DiffNet` constructs a summary within a reasonable time frame.

We further evaluate the running time of `DiffNet` at varying network density. Figure 6.11c shows the running time on [6] dataset network from 10% density (0.1) to

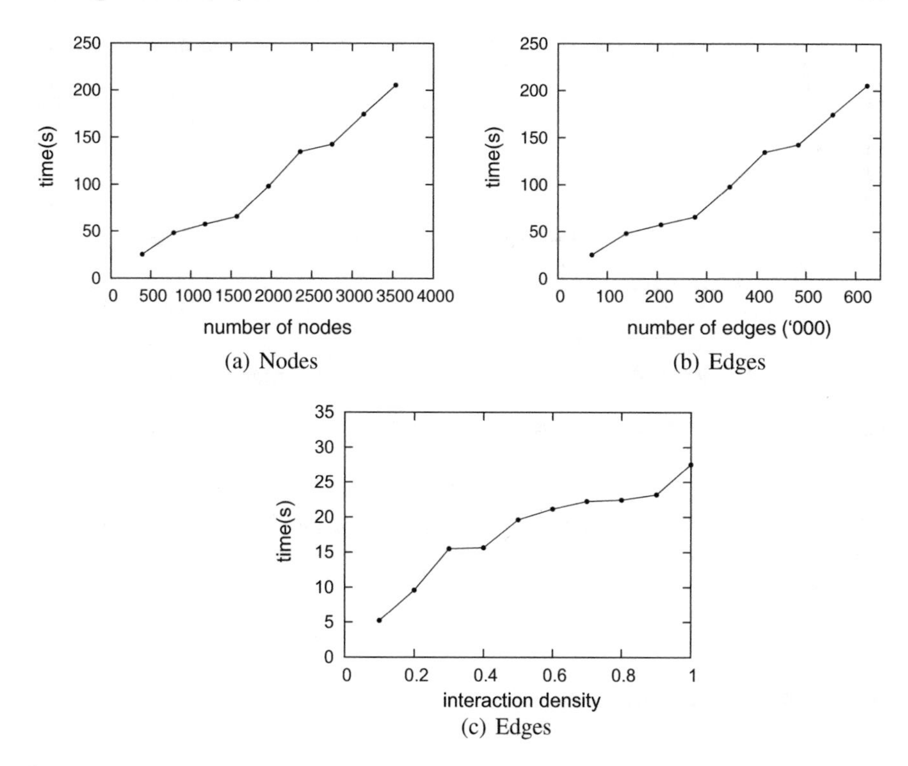

Fig. 6.11 Running time of `DiffNet`

full density (1.0). We artificially construct networks at varying density by randomly removing network edges until the desired density is achieved. From the figure, running time of `DiffNet` grows almost linearly with the network density.

6.6.5 *Effect of Interaction Noise*

Given that interaction profiles are likely to be noisy, we evaluate the effect of interaction noise on `DiffNet` summary construction. We assume the `DiffNet` summary generated from the differential network in [6] is without interaction noise and use it as the reference summary. We then simulate the effect of noise by perturbing the interactions of the network by random rearrangement of its interactions. The amount of perturbation is indicated by the *interaction noise rate*, which is the fraction of the original interactions that have been randomly rearranged. Figure 6.12a shows the stability of of the `DiffNet` summary after interaction noise perturbation. At each noise rate, we simulate 10 perturbed network samples. We compute the Jaccard similarity of the functional subgraphs of a perturbed summary ($\mathscr{S}_1$) against the reference summary ($\mathscr{S}_2$). Specifically $JaccardSimilarity(\mathscr{S}_1, \mathscr{S}_2) = 1$ if the gene

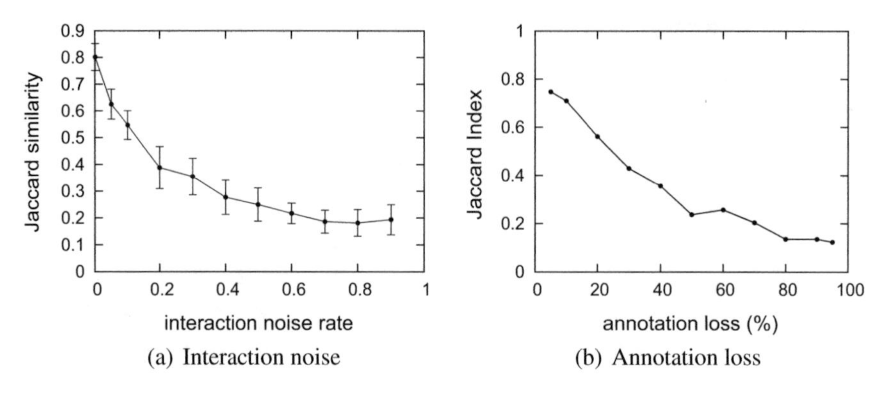

(a) Interaction noise (b) Annotation loss

Fig. 6.12 Effect of interaction noise and loss of annotations

set of each functional subgraph in $\mathcal{S}_1$ and $\mathcal{S}_2$ is identical. As expected, we observe a steady decrease in similarity against the reference summary with increasing interaction noise rate.

6.6.6 *Effect of Annotation Loss*

As current gene annotations are likely to be incomplete, here we study the effect of gradually removing gene annotations on `DiffNet` summary construction.

Suppose $\mathcal{S}_0$ is a reference `DiffNet` summary of the [6] differential network with complete gene annotations. We then construct `DiffNet` summaries of differential networks with removed annotations and observe their similarities with the reference summary. Given two summaries $\mathcal{S}_1$ and $\mathcal{S}_2$, the similarity of the functional subgraphs between the summaries can be measured using the following:

$$JaccardIndex(\mathcal{S}_1, \mathcal{S}_2) = \frac{1}{|\mathcal{S}_1|} \sum_{C_1 \in \mathcal{S}_1} \max_{C_2 \in \mathcal{S}_2} \frac{|V_1 \cap V_2|}{|V_1 \cup V_2|} \tag{6.9}$$

where $JaccardIndex(\mathcal{S}_1, \mathcal{S}_2) = 1$ if the gene set of each functional subgraph in $\mathcal{S}_1$ and $\mathcal{S}_2$ is identical. We remove $n\%$ of the gene annotations from the differential network and construct a new summary $\mathcal{S}_n$. We call $\mathcal{S}_n$ a summary of the differential network with *n% annotation loss*. Figure 6.12b shows the $JaccardIndex$ similarities of summaries with varying annotation loss. We observe that annotation loss creates a summary that is increasingly different from the reference summary. The drop in $JaccardIndex$ similarity is gradual, suggesting that `DiffNet` summary construction is relatively robust to annotation noise. More importantly, as annotations of genes are likely to increase with time, it will only lead to more improved performance of `DiffNet`.

6.7 Conclusions

In this chapter, we present DiffNet, a data-driven algorithm that automatically constructs summaries of differential functional responses of gene interaction networks under environment or condition change. Specifically, it leverages combination of GO annotation information and underlying interaction data to greedily identify a set of functional subgraphs that are highly skewed and coherent, representing significant functional responses due to condition change. Experimental study with a real-world network revealed that DiffNet can automatically generate high quality differential functional summaries from the differential network including differential interactions that [6] failed to identify. Furthermore, state-of-the-art graph clustering algorithms cannot be adopted to generate such differential summaries. In fact, by incorporating the notion of coherence and skewness, DiffNet is able identify key functions that coherently respond to condition change whereas traditional clustering methods fail to do so. Lastly, DiffNet is efficient and can generate differential summaries in acceptable time.

References

1. B.S. Seah, S.S. Bhowmick, C.F. Dewey Jr., Diffnet: automatic differential functional summarization of de-map networks. Methods **69**, 247–256 (2014)
2. S.R. Collins, M. Schuldiner, N.J. Krogan, J.S. Weissman, A strategy for extracting and analyzing large-scale quantitative epistatic interaction data. Gen. Biol. **7**, R63 (2006)
3. R.P. St Onge, R. Mani, J. Oh, M. Proctor, E. Fung, R.W. Davis, C. Nislow, F.P. Roth, G. Giaever, Systematic pathway analysis using high-resolution fitness profiling of combinatorial gene deletions. Nat. Genet. **39**, 199–206 (2007)
4. T.M. Przytycka, M. Singh, D.K. Slonim, Toward the dynamic interactome: it's about time. Brief. Bioinf. **11**, 15–29 (2010)
5. T. Ideker, N.J. Krogan, Differential network biology. Mol. Syst. Biol. **8**, 565 (2012)
6. S. Bandyopadhyay, M. Mehta, D. Kuo, M.-K. Sung, R. Chuang, E.J. Jaehnig, B. Bodenmiller, K. Licon, W. Copeland, M. Shales, D. Fiedler, J. Dutkowski, A. Guénolé, H. van Attikum, K.M. Shokat, R.D. Kolodner, W.-K. Huh, R. Aebersold, M.-C. Keogh, N.J. Krogan, T. Ideker, Rewiring of genetic networks in responseto DNA damage. Science (New York, N.Y.) **330**, 1385–1389 (2010)
7. M. Schuldiner, S.R. Collins, J.S. Weissman, N.J. Krogan, Quantitative genetic analysis in Saccharomyces cerevisiae using epistatic miniarray profiles (E-MAPs) and its application to chromatin functions. Methods (San Diego, Calif.) **40**, 344–352 (2006)
8. G.D. Bader, C.W.V. Hogue, An automated method for finding molecular complexes in large protein interaction networks. BMC Bioinf. **4**, 2 (2003)
9. B.J. Frey, D. Dueck, Clustering by passing messages between data points. Science (New York, N.Y.) **315**, 972–976 (2007)
10. T. Nepusz, H. Yu, A. Paccanaro, Detecting overlapping protein complexes in protein-protein interaction networks. Nat. Methods **9**, 471–472 (2012)
11. E.I. Boyle, S. Weng, J. Gollub, H. Jin, D. Botstein, J.M. Cherry, G. Sherlock, GO::termfinder–open source software for accessing Gene Ontology information and finding significantly enriched Gene Ontology terms associated with a list of genes. Bioinformatics (Oxford, England) **20**, 3710–3715 (2004)

12. A. Subramanian, P. Tamayo, V.K. Mootha, S. Mukherjee, B.L. Ebert, M.A. Gillette, A. Paulovich, S.L. Pomeroy, T.R. Golub, E.S. Lander, J.P. Mesirov, Gene set enrichment analysis: aknowledge-based approach for interpreting genome-wide expression profiles. in *Proceedings of the National Academy of Sciences of the United States of America*, vol. 102 (2005), pp. 15545–15550
13. A. Subramanian, H. Kuehn, J. Gould, P. Tamayo, J.P. Mesirov, GSEA-P: a desktop application for gene set enrichment analysis. Bioinformatics (Oxford, England) **23**, 3251–3253 (2007)
14. J. Gillis, M. Mistry, P. Pavlidis, Gene function analysis incomplex data sets using ErmineJ. Nat. Protoc. **5**, 1148–1159 (2010)
15. A. Jain, K. Mohiuddin, Artificial neural networks: a tutorial. Computer **29**, 31–44 (1996)
16. V. Chvatal, A greedy heuristic for the set-covering problem. Math. Oper. Res. **4**, 233–235 (1979)
17. P. Fabrizio, F. Pozza, S.D. Pletcher, C.M. Gendron, V.D. Longo, Regulation of longevity and stress resistance by Sch9 in yeast. Science (New York, N.Y.) **292**, 288–290 (2001)
18. H. Rodriguez-Rocha, A. Garcia-Garcia, M.I. Panayiotidis, R. Franco, DNA damage and autophagy. Mutat Res. **711**(1–2), 158–166 (2011)
19. B.E. Kremer, L.A. Adang, I.G. Macara, Septins regulate actin organization and cell-cycle arrest through nuclear accumulation of NCK mediated by SOCS7. Cell **130**(5), 837–850 (2007)
20. A.J. Enright, S. Van Dongen, C.A. Ouzounis, An efficient algorithm for large-scale detection of protein families. Nucl. Acids Res. **30**, 1575–1584 (2002)
21. B. Zhang, B.-H. Park, T. Karpinets, N.F. Samatova, From pull-down data to protein interaction networks and complexes with biological relevance. Bioinformatics (Oxford, England) **24**, 979–986 (2008)

Chapter 7
The Road Ahead

In this chapter, we summarize the contributions of this book and establish several lines of inquiry associated with clustering and summarizing biological networks for future research.

7.1 Summary

The contributions of this book are summarized as follows:

- In Chap. 3, we discussed efforts by the bioinformatics and data mining communities on summarizing the functional organization within a PPI network by network clustering. Our approach has been to emphasis the unique characteristics of the network clustering problem in the context of PPI networks and discuss an array of techniques highlighting their strengths and limitations. This analysis is fundamental in developing effective computational solutions toward comprehending the organization and functioning of cells.
- In Chap. 4, we present a data-driven and generic PPI network summarization framework called FUSE to generating functional summaries at multiple resolutions from a PPI to provide a high level view of its functional landscape. It constructs higher level functional summary that summarizes the underlying PPI network to obtain a concise, interpretable representation of the network. It generates the "best" summary from both interaction and annotation data by maximizing information gain for a specific resolution. We demonstrate the role of FUSE in addressing the information overload issue of analyzing large scale PPI networks. We evaluate the performance of FUSE on several real-world PPIs. We also compare FUSE to state-of-the-art graph clustering methods with GO term enrichment by constructing the biological process landscape of the PPIs. Our experimental results demonstrate that FUSE is highly effective in constructing higher order functional maps with

S.S. Bhowmick and B.-S. Seah, *Summarizing Biological Networks*, Computational Biology 24, DOI 10.1007/978-3-319-54621-6_7 superior accuracy and representativeness compared to these state-of-the-art graph clustering methods. Using Alzheimer's Disease network as our case study, we further demonstrate the ability of FUSE to quickly summarize the network and identify many different processes and complexes that regulate it. We analyze the topological features of the functional landscape of human PPI that leads us to the identification of *functional hubs* (clusters of proteins that act as hubs).

- In Chap. 5, we present FACETS, a data-driven and generic algorithm for generating multi-faceted functional summarization of a PPI network, providing multiple perspectives of the functional organization landscape of the network. Each perspective (facet) in the atlas represents a distinct interpretation of how the network can be functionally summarized. Specifically, FACETS maximizes interpretative value of the atlas by optimizing inter-facet orthogonality and intra-facet cluster modularity. The performance of FACETS was extensively discussed with several real-world PPI networks. We also performed a case study using human autophagy system to illustrate the utility of this framework. In summary, the experimental analysis demonstrates the effectiveness of FACETS in generating functionally distinctive facets. These distinctive facets have higher relevance to real life datasets compared to single decomposition-based graph clustering techniques.
- Finally, in Chap. 6 we present DiffNet, a data-driven technique that automatically constructs summaries of differential functional responses of gene interaction networks under environment or condition change. Specifically, it leverages combination of GO annotation information and underlying interaction data to greedily identify a set of functional subgraphs that are highly skewed and coherent, representing significant functional responses due to condition change. DiffNet is one of the first step towards generating differential functional summary between two snapshots of E-MAP networks under contrasting environmental conditions. Our exhaustive experimental evaluation with real-world dataset demonstrates the superiority of this technique to address the differential network summarization problem.

7.2 Future Research

In general, network clustering and summarization approaches have largely been to limited to static PPI networks. Where previously biologists have concentrated on large scale static networks, there is now a surge of interest in constructing large scale quantitative "omics" data, including quantitative PPI and gene interaction networks. In light of this, we suggest several future research in this direction as follows.

Superior clustering techniques. Recall from Chap. 3, a desirable PPI network clustering algorithm needs to cover the entire network but at the same time ensure that it is scalable, exhaustive, robust against noisy edges, admits overlapping clusters, and leverages on annotations whenever possible. While existing techniques support a subset of these features, realization of a clustering approach that supports all these features effectively remains an open research problem. Furthermore, how to accurately measure the quality of predicted clusters in the presence of noisy edges is still a challenging problem.

Enabling discovery of disease modules. Existing clustering techniques largely focus on detecting topological and functional modules. Recent research have demonstrated that a group of proteins involved in a disease demonstrates a high inclination to interact with each other forming a *disease module* [1]. Note that a disease module may not necessarily be same as a topological or functional module. However, it may overlap with the latter modules. A disease module is defined in the context of a specific disease and each disease is associated with a unique module. Knowledge of these disease modules can pave way for network-based disease gene prediction during drug development. Intuitively, if a few proteins of a disease module are unveiled, then there is a high possibility that other disease-related proteins can be discovered in their vicinity. This can play a pivotal role in drug discovery as many existing drugs are palliative in nature (i.e., they target proteins in the network neighborhood of a disease-related protein instead of the latter) [2]. It is interesting to explore how PPI clustering techniques can enable us to discover these disease modules.

Comprehensive summarization techniques. As the network summarization techniques discussed in this book rely on functional information, they are unable to summarize regions without functional information. This could be important when there is a species network with many proteins of unknown functions. To address this, one direction of work is to extend these summarization algorithms to handle such regions. Another direction will be to incorporate genomic and proteomic experimental data to enhance the quality and comprehensiveness of summaries.

Quantitative network summarization. In Chap. 4, we discussed FUSE, an algorithm for summarizing PPI networks. PPI networks are static and limited in its power to model the complex behavior of the biological system. In fact, biological systems cannot be accurately modeled as static networks; they are dynamic and respond to both environmental and genetic factors. Therefore, quantitative models that can incorporate the dynamic properties of biological systems are increasingly important. Among existing quantitative models are *ordinary differential equations* (ODEs) and *partial differential equations* (PDEs) models. To this end, we presented the `DiffNet` method as a first step towards quantitative network summarization (Chap. 6). However, `DiffNet` is limited to binary snapshots of the network and cannot easily be extend to systems that model a continuum of states (e.g., ODE and PDE models). As part of future research, there is opportunity to extend the notion of quantitative network summarization to more powerful ODE and PDE quantitative models. The complexity associated with understanding ODE and PDE models is well known, and extending network summarization to such quantitative models may enable us to better visualize their dynamic behavior in a multi-perspective manner. This is a challenging problem that requires careful investigation.

References

1. A.L. Barabasi, N. Gulbahce, J. Loscalzo, Network medicine: a network-based approach to human disease. Nat. Rev. Genet. **12**, 56–68 (2011)
2. M.A. Yildrim, K.-I. Goh et al., Drug-target network. Nat. Biotechnol. **25**, 1119–1126 (2007)

Index

Note: Page numbers followed by f and t indicate figures and tables respectively

S.S. Bhowmick and B.-S. Seah, *Summarizing Biological Networks*, Computational Biology 24, DOI 10.1007/978-3-319-54621-6

Printed in the United States
By Bookmasters